MICROPROCESSOR ARCHITECTURE
AND PROGRAMMING

Microprocessor Architecture and Programming

William F. Leahy

Chairman—School of Engineering
Pacific States University
Los Angeles, California

Member of the Technical Staff
Rockwell International
Anaheim, California

A Wiley-Interscience Publication
JOHN WILEY & SONS
New York • London • Sydney • Toronto

Library of Congress Cataloging in Publication Data:

Leahy, William F.
    Microprocessor architecture and programming.

    "A Wiley-Interscience publication."
    Includes bibliographical references and index.
    1. Microprocessors.   2. Computer architecture.
I. Title.

QA76.5.L377   001.6'4'04   77-1552
ISBN 0-471-01889-9

Printed in the United States of America

10 9 8 7 6 5 4 3 2

To Marianne, William Randolph, and Randene

PREFACE

The age of the microprocessor has introduced a technology revolution in the field of electronics that is destined to rival and surpass the transformations caused by the introduction of the bipolar transistor in the 1950s. This technology is not only affecting the technologically oriented professional (the electronic engineer, programmer, and scientist) but is also creating new business opportunities in consumer electronics by opening the doors of complex "smart" electronic systems to the small-business entrepreneur.

The elaborate "space age" computer-controlled systems, which once cost many thousands of dollars and required gigantic engineering conglomerates to design and build, are today rapidly being supplanted by the use of dedicated microprocessor controllers at a fraction of the cost. Many of these systems are designed and built by electronic newcomers, whose concerns are barely larger than the proverbial "kitchen table business." Applications that were exempt from computer control because of exorbitant cost are now prime candidates for reduced cost, both in initial investment and cost of ownership, through the utilization of the microcontroller. The businessman or professional who needs to "stay on top" of his field must be knowledgeable of this emerging field or relinquish his position to the professional who is.

With the introduction of the microprocessor, the traditional tasks of the analog and digital engineer have also been modified. Vast areas of design technology that were once reserved for the analog design engineer are now being performed with greater accuracy, higher reliability, and lower cost by the use of the digital microprocessor. In like fashion, the field of digital technology is being transformed from the use of discrete small-scale integrated components to the use of microprocessor systems. The entire field of electronic design engineering is undergoing a major transformation that the engineer must rapidly adapt to or become obsolete.

In short, every field of experience is being affected by the microprocessor revolution. No professional, whether in management, marketing, or electronic design, will be able to maintain his position in the world of technology unless he gains a working knowledge of the microprocessor.

This text is intended to aid in the educational transition process from the sphere of discrete electronic technologies to the medium- and large-scale integration techniques used in the microprocessor field. The business manager or design engineer who must weigh the cost of advanced technologies against the increased performance and marketability of his product will find insight for these decisions in Chapters 1 through 3. The unique components used in the microprocessor system are described in Chapters 4 through 9, with particular emphasis placed on the design of the microprocessor memory systems, since the memory usually accounts for the largest share of the microcontroller cost.

Once the reader understands the hardware requirements of the microprocessor described in Chapters 1 through 9, he can turn to the detailed study of machine-level programming presented in Chapters 10 through 12 and the extensive programming illustrations provided in Chapters 13 and 14. This study utilizes the Intel 8008 microprocessor machine language.

The text concludes with the application of the programming techniques to the Intel 8080 and Motorola M6800 microprocessor chips (Chapter 15). These devices represent two of the more popular second-generation microprocessor chips on the market.

WILLIAM F. LEAHY

Yorba Linda, California
January 1977

CONTENTS

MICROPROCESSOR ARCHITECTURE
AND PROGRAMMING

CHAPTER 1

Introduction

The subject of microprocessors has evoked many responses from the practicing electronic engineer, ranging from amazement and wonder to passive aloofness and from excitement to disdain. No other engineering component introduced within the last two decades has been greeted with such obviously mixed feelings or misunderstanding by the engineering community.

The main ingredient for this misunderstanding is, in the opinion of the author, the manner in which the microprocessor has been introduced to the engineering community by the device vendors. With its introduction in the engineering literature in the early 1970s, it fell into the sphere of neither a component nor a system; it was neither a tool nor a solution. The component/circuit engineer was unwilling to utilize the microprocessor since he viewed it as a system, not an electronic component/circuit. The system engineer refused to use it as it did not perform a function in itself and, therefore, required the component design engineer to transform it into a usable piece of hardware. The engineering programmer avoided its use since, unlike the traditional computer, it required significantly more than programming to utilize its power. In short, the traditional division of design engineering responsibilities rendered the microprocessor an "orphan" component cutting across the engineering disciplines without being accepted by any.

In an attempt to alleviate this awkward position, multiple seminar courses have been offered to the engineer at large by vendors of microprocessing equipment as well as private educational organizations. In general, these efforts again suffered from a lack of a clear definition of the engineering disciplines that are in existence today and will remain for the foreseeable future. It is, in general, an unviable assumption that the design engineer will desire to be transformed into a computer programmer or the engineering programmer into an electronic designer in order to utilize a microprocessing system. Yet, the microprocessor stands as a tool desperately needed by both the designer and the programmer to provide dynamic system solutions to the engineering problems of the twenty-first century.

It is the attempt of this treatment of the microprocessor art to place the microprocessor technology into perspective within the design engineering discipline and to indicate its potential as an engineering component/circuit design tool. Attention is placed on the programming of the microprocessor as it applies to ensuring that the system performs its design function, while the intricacies of programming are best left to the programming disciplines.

Particular emphasis is placed here on the interrelated technical and economic trade-offs of memory, interface electronics, and central processing unit (CPU) programming capabilities as they pertain to the design of the microprocessing system.

It should be noted that there remains a strong attraction to transform this treatment of the microprocessor technology into a compendium of the microprocessor systems/devices flooding the market at the present time. Such an effort would prove fruitless, however, as many of the devices presently on the market will ultimately be proven obsolete as technology advances or undesirable because of poor architecture. No treatment of a subject as complex as the microprocessor, however, can be presented in a vacuum of generalities and still remain useful to the engineer. For this reason, the selection of the first marketed 8-bit microprocessor (Intel 8008) as the *model* device for evaluation is both a natural and desirable selection.

The Intel 8008 is a natural selection for the *model* since it was used to initiate the microprocessor form as we know it today. It served as the first approach to microprocessor architecture, providing a test bed to determine both the adequate and not-so-adequate features of this first concept. The myriad devices that have since followed include the Intel 8080 second-generation processor (see Chapter 15) as well as chips by National Semiconductor, Motorola, Signetics, and others. While these later chips are vastly more powerful in their instruction and memory capabilities, they are generally more system oriented, yielding less insight into the basic makeup of the microprocessor proper. It is believed that the transition from the basic Intel 8008 processor architecture to the more complex systems available today will be minor once the basic microprocessor language and operation is comprehended.

The choice of the Intel 8008 is also a desirable selection since the chip is readily available in discount electronic stores at present, is inexpensive (less than $20), can operate with virtually any readily available memory chip and, therefore, is obtainable to the student in a hands-on learning situation, in either the manufacturing or home laboratory. There is no learning situation to compare with that of physical learning by hands-on experience. Textbook education is *cryptic,* becoming transformed into knowledge by the physical application of the *cryptic* to the pragmatic hands-on operating circuit. While exotic new microprocessor systems are being presented to the engineering community virtually on a daily basis, few systems have become "off-the-shelf" deliverable units. Thus, while many will criticize the Intel 8008 system for its numerous shortcomings, no one can complain that it is not readily available.

CHAPTER 2

The Microprocessor—An Overview

The microprocessor is an integrated computer that has been shrunk down to a "micro"/miniature size. In an oversimplification of the concept, the microprocessor is a "computer on a chip."

To understand the operation and capabilities of the microprocessor, it is necessary first to gain an understanding of basic computer architecture and to apply this knowledge to the microprocessor technology as it has developed today. A computer in this context is defined as any device capable of automatically carrying out a sequence of operations on data that are either digital or analog in form. The operations referred to may be simple computational operations (add, subtract, multiply, divide, etc.) or logic-decision operations (AND, OR, equality, etc.). The results of these operations are then combined to develop complex computational calculations for direct on-line control processes or indirect analytical analysis.

BASIC ELEMENTS OF THE CLASSICAL COMPUTER

The architecture of the classical computer is centered around four basic building blocks or units. These elements, shown in Figure 2.1, are

1. The **arithmetic/logic unit (ALU),** which is frequently referred to as the central processing unit (CPU)
2. The **control unit**
3. The **memory unit,** which contains both the program memory (PM) and temporary storage memory
4. The **input/output unit**

To aid in understanding the operation and interrelationship of each of the four building blocks, an analysis of the fundamental task of each block is performed.

The Arithmetic/Logic Unit

The ALU is the "muscle" of any computer, performing the basic mathematical and logic function manipulations necessary for decisionmaking and analytical computation. The ALU, as such, defines the basic capability of the machine.

The typical ALU is designed to perform multiple arithmetic and logic functions including

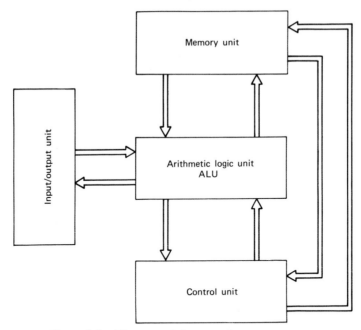

Figure 2.1. Elements of the general-purpose computer.

1. **ADD** two numbers

2. **SUBTRACT** one number from another

3. **SHIFT** a number to the RIGHT by one bit

4. **SHIFT** a number to the LEFT by one bit

5. **COMPARE** two numbers and identify **equality**

6. Logically **AND** two numbers together

7. Logically **OR** two numbers together

8. Logically **NAND** two numbers together

9. Logically **NOR** two numbers together

10. Logically perform **INCLUSIVE** or **EXCLUSIVE AND/OR** functions on two numbers

It should be emphasized that the ALU does not have the capability of multiplication, division, or performing other complex mathematical functions beyond simple addition and subtraction. The capability of performing complex mathematical functions, as well as high-complexity decisionmaking, is a direct function of computer programming. The computer program (called an *algorithm*) defines how to combine and sequence the simple ALU functions in order to form the complex mathematical and logical manipulations which have come to be synonymous with the modern-day computer. The ALU is, therefore, a robot of "limited capability" that gains its reputation from the creative power introduced to it by means of the program-control unit.

The Program-control Unit

The program-control unit ultimately controls all functions of the computer and directs all of its operations; in short, it acts as brain center and nervous system of the computer. To perform this major task the program control unit can be subdivided into four basic units. They are

1. The master timing unit
2. The program storage memory
3. The stack pointer memory unit
4. The internal program switching matrix

Figure 2.2 illustrates the interrelationship of these four subelements.

Master Timing Unit

The master timing unit generates and controls the basic timing sequences for the computer. This control is centered on the time sequence generated by the master control cycle, which is continuously repeated within the computer unless commanded to stop either by program control (a halt signal is received) or by means of an external operator-issued command. A typical four-step master machine timing cycle is illustrated in Figure 2.3.

The master control timing cycle is organized as follows:

Minor cycle 1. The program control counter (PCC) value is shifted into the memory address register (MAR). This value defines the memory address location of the computer instruction that is to be performed during the current machine cycle.

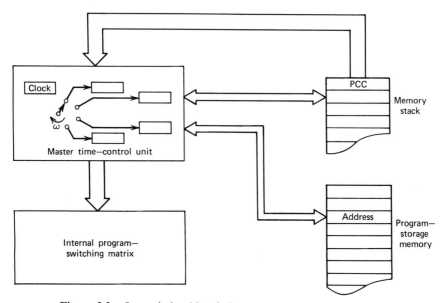

Figure 2.2. Interrelationship of elements of program-control unit.

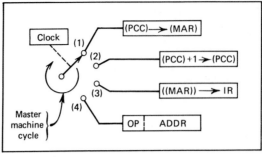

PCC ≡ Program—control counter
MAR ≡ Memory—address register
((MAR)) ≡ Contents of memory addressed
IR ≡ Instruction register

Figure 2.3.   Four-step master machine timing cycle.

Minor cycle 2. The PCC is incremented by one count in preparation for the next machine cycle. By this technique the entire program is incrementally stepped through by the computer control, usually being initiated at 00 and stepping one count at a time through the entire predefined program.

Minor cycle 3. The instruction located in the address location established during the first minor cycle is transferred to the instruction register. The instruction register controls the states of the operational switches of the computer. These operational switches direct data transfer between the elements of the computer and establish the arithmetic/logic function to be performed by the ALU.

Minor cycle 4. The instruction register is "read" and the appropriate operational switches are physically opened or closed to perform the designated instruction operation. It is at this point that the data are operated on and the desired instruction performed.

The master machine-timing cycle is now complete. If the instruction performed was not a *halt* instruction, the control unit would initiate another machine cycle composed of the same four minor cycles but beginning with the PCC at a value one count higher than on the previous cycle.

The exact timing of the master control timing cycle with its associated four minor cycles is controlled by the computer clock, which is derived from the master oscillator. Acting like a constant-speed rotating switch, the master control timing cycle is swept out in a constantly repeating time sequence.

Program-storage Memory

The program-storage memory is the primary element controlled by the program-control unit. The program-storage memory maintains the exact sequence of events (called the *computer program*) that the computer performs.

During the first minor cycle the PCC establishes which address location within the program-storage memory the control unit will "read." Once this address location is "read" the contents of this address (called the *operation code*) is transferred into the instruction register. During the fourth minor cycle the operation code corresponding to the data placed in the instruction register sets up the computer's operational switches, which result in the issuance of the appropriate machine commands to the other elements of the computer. These commands are performed during the fourth minor cycle.

The program storage memory has been referred to as the "brain" of the computer since it details the complete program that the unit will follow. This memory can be: (a) temporary, modifiable by the programmer just prior to performing a computer run (in the case of the general-purpose computer) by means of providing a new sequence of punched cards to be read into the program memory or (b) permanent (in the case of a dedicated microprocessor), in which there can be no modification of the program steps without minor component changes. The programming of this memory is examined in Chapters 6, 7, and 10.

Stack-pointer Memory Unit

Closely associated with the program counter and program-memory unit is the stack-pointer memory unit. The stack-pointer memory is composed of a stack of memory locations, the top location being occupied by the present program counter. During normal operation the program counter is incremented by one count with every master timing cycle, the stack appearing to have no depth. However, the genius of the computer lies in its ability to change its program sequence (to perform decisionmaking) based on external or internal conditions. The stack depth is used to accommodate this change of program sequence.

During a typical program sequence the program counter is incremented by one count with each master timing cycle. In this way the program storage memory is stepped through in sequence, each memory address being selected and read in exact order. However, should the program require a modification of this sequence (a transfer to a different program address) as a result of a decision step, the program counter is modified to correspond to the initial location of the new program sequence. To accomplish this program counter modification the existing program counter value is usually stored (pushed down) within the stack while the new value of the program counter (address of the new program sequence) is entered into the top location of the stack. During the next master timing cycle the new program counter is incremented by one count to step through the new program memory sequence while the old program counter (one level down in the stack) remains at its last count value (address).

This technique of modifying the program sequence by pushing down the stack can continue until the stack depth has been reached, permitting multiple decisions to be made simultaneously by the computer. Each time a new sequence is called as a result of a decision, the stack is pushed down one level. This technique is called *nesting*. If the stack depth is exceeded, the original program counter value is lost, and could cause the overall program to fail.

In general, a transfer to another program sequence is intended not as a permanent program modification, but is rather a temporary modification with the intent to return to the main program once the new sequence is completed. To accomplish the return from the modified program to the main program, a **RETURN** instruc-

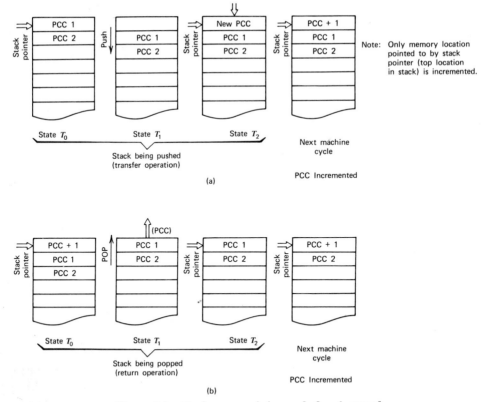

Figure 2.4. Stack memory being **pushed** and **popped**.

tion is inserted into the memory program. This operation code instructs the stack pointer memory to reestablish the last program counter value nested to the main program counter (the top of the stack). The stack is then pushed up or *popped,* returning the program control to the previous program control counter value. By the use of as many **RETURN** program statements as program decisions, the stack is popped as often as it was pushed and the original program control counter is returned to the top position in the stack. Figure 2.4*a* illustrates the technique of pushing down a stack, while Figure 2.4*b* illustrates popping the stack.

Internal Program Switching Matrix

The internal program switching matrix (operational switches) is the traffic controller for the computer. All other portions of the program-control unit function to establish the sequence of switch closures and openings within the internal device switching matrix to perform the various operations associated with the computer. The actual data flow is also conducted through this switching matrix to the internal and external elements of the computer.

The complexity of this matrix extends the power of the computer beyond the limited functions of the ALU to include data-transfer functions, such as moving data from the accumulator to other temporary registers or main memory, or vice versa. The matrix also controls the input/output units, controlling the flow of data between the internal operations of the computer and the external environment.

The Main Memory

The largest major internal element of the computer architecture is the memory unit. The memory unit has only two functions: (a) to house the program storage memory holding the operation instructions used by the program control unit and (b) to store data.

The program storage memory can be either read/write memory or read-only memory, depending on the design of the computer. In the microprocessor applications this portion of memory is generally composed of permanent read-only memory. The portion of memory devoted to data storage is always read/write memory and is used to store: (a) input data for future use, (b) intermediate calculations, or (c) final results prior to their output to an interface equipment.

In most computers the memory comprises the bulk of the electronics making up the system. This is due to the centrality of the operating ALU and program-control unit and the diversity of program-storage memory and data subsets to be stored, operated on, and the final results restored. Thus, while the overall use of the memory is simple, it stands as the primary hardware component in the computer.

The Input/Output Unit

The input/output unit of the computer constitutes the man–machine or machine–machine interface. In general, the image of the typewriter or numerical keyboard is evoked when thinking of computer input units and printers or television screens visualized when thinking of computer output units. These units are good examples of the man–machine input and output interface but are not representative of the endless possibilities of input/output units that constitute the machine–machine interface. These input units include temperature and pressure sensors, photoelectric cells, electromechanical position indicators, and other physical parameter sensors. All of these sensors must, of course, transform their measured parameters into a digital format that is compatible with the digital computer language. The associated output units for machine–machine interface are composed of motors, solenoids, and electromechanical valves.

In every case cited, however, the input/output unit is actually a peripheral piece of hardware interfacing with the input or output register of the computer. These registers constitute the input/output of the computer, since any piece of peripheral equipment supplying the proper digital interface can be used at the input/output unit interface register.

The input/output registers are temporary storage registers capable of inputting data from the peripheral equipment (**INPUT**) or from the computer (**OUTPUT**) and having these data locked in until called for. Once the information is accepted by the computer (**INPUT**) or peripheral equipment (**OUTPUT**) the register is unlocked and readied for the next set of data received.

THE COMPUTER IN OPERATION

To illustrate the operation of the classical computer configured as shown in Figure 2.5, compare two systems for solving a simple addition of two numbers. The first system is a man/hand-held calculator computer system. The second system is a fully automated "microprocessor" system.

For the man–calculator system, the memory system (both program memory and temporary storage memory) is a piece of paper. For program storage, a list of printed instructions are provided to direct the man in the proper operation of the calculator to reach the solution to the problem. The problem to solve for X in $6 + 8 = X$.

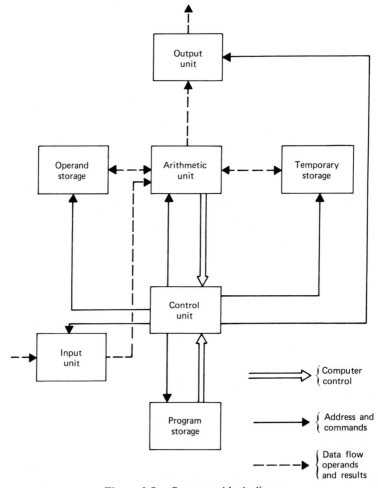

Figure 2.5. Computer block diagram.

The man–calculator program might be performed using the following procedure:

Detailed Program		System Operation
1. Clear calculator.	1 cycle	1. Read: operator reads first program step.
		2. Execute: operator presses "clear" key on calculator.
2. Enter number *A* into calculator.	1 cycle	3. Read: operator reads second program step.
		4. Fetch: operator looks into memory (the paper) to find value of number *A*—(6).
		5. Execute: operator presses "6" key on calculator.
3. Press addition function (+) key.	1 cycle	6. Read: operator reads third program step.
		7. Execute: operator presses addition function key.
4. Enter number *B* into calculator.	1 cycle	8. Read: operator reads fourth program step.
		9. Fetch: operator looks into memory (the paper) to find value of number *B*—(8).
		10. Execute: operator presses "8" key on calculator.
5. Press equality function (=) key.	1 cycle	11. Read: operator reads fifth program step.
		12. Execute: operator presses equality function (=) key.
6. Read and record answer.	1 cycle	13. Read: operator reads sixth program step.
		14. Execute: operator reads answer (14) and records it on paper.
7. Halt.	1 cycle	15. Read: operator reads seventh program step.
		16. Execute: operator stops.

The program has fully been executed in 16 steps, seven cycles. (Note that a cycle, the time duration necessary to perform a basic flow chart command, varies in duration dependent upon the command desired.) The cycle is a normal measure of machine timing and, as a minimum, should be composed of a read/execute instruction set. It can, however, be much longer than a minimum two instruction set.

When this basic problem is presented to the computer system rather than a man–calculator system, an expanded program must be provided (i.e., all operations must be uniquely specified since the computer has no capability for independent thought) to permit the control unit to direct the overall sequence of the computer system. This program is located in the program memory. The step number that the computer is performing at a given time is maintained by the program counter. In addition, an input/output peripheral device (typewriter) is provided to interface with the computer arithmetic system in order to input the data and read out the resultant answer.

Figure 2.6 illustrates the minimum logic interconnections between the four main computer elements required to perform this simple addition function. The switches

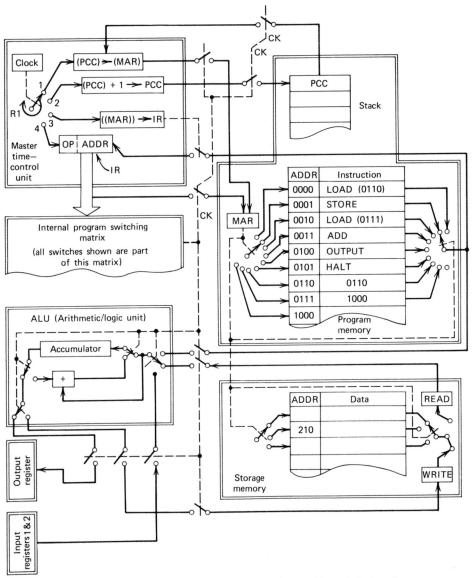

Figure 2.6. System interconnection to solve problem 6 + 8 = ?

shown in the diagram are controlled by the operational code being performed and are part of the internal program switching matrix.

It is useful to describe some of the computer operations by means of a short-hand notation that is applicable to programming of both the large computer and the smaller microprocessor. Single parentheses around a register designation denote "the contents of that register." An arrow between two expressions indicates the movement of data between registers in the direction of the arrow. Double parentheses denote the contents of the memory location to which the contents of the designated memory address are pointing. Table 2.1 illustrates this shorthand notation. With an expanded computer system, the machine–machine level program could be written as follows.

Instruction Memory Location—PCC	Instruction	System Operation
0000	Input first data number to temporary storage—accumulator (A).	The number (6) at the input terminals is transferred to the CPU accumulator memory. This location is generally used as the location for temporary storage of both input and output data.
0001	Transfer accumulator to memory location 210.	The number (6) is transferred from the accumulator to memory location 210.
0002	Input second data number to temporary storage (accumulator).	The number (8) at the input terminals is transferred to the CPU accumulator.
0003	Add memory location 201 to accumulator.	The number in memory location 201 (6) is added to the number presently in the accumulator (8) and the result (14) is left in the accumulator.
0004	Output accumulator.	The number (14) in the accumulator is outputted to the output unit.
0005	Halt.	The computer system stops.

The machine-level cycle sequence to perform this program is shown in Table 2.2. The program-control counter initiates at number 0000, the processor calls up the program instruction stored in the program memory at location 0000, decodes the operation (input data), and executes it by means of control signals issued to the input unit. On completion of the instruction located at 0000, the program control counter, having been incremented by one count (to 0001), calls up the instruction located at 0001, decodes it, and commands the appropriate command signals to perform the indicated instruction. This continues until the program-control counter reaches instruction 0005, at which time the computer system stops. It should be noted that the act of turning on the computer system forces the program control counter to the first instruction location, which is location 0000 in this illustration. This initialization is often called the *computer master reset command.*

TABLE 2.1. Computer Shorthand Notation

Notation	Meaning
(PCC)	The contents of the program control counter.
(PCC)→ MAR	The contents of the program control counter are moved to the memory address register.
((MAR))→ IR	The contents of the memory location pointed to by the memory address register are moved to the instruction register.

TABLE 2.2. Machine Level Cycle Sequence

Master Cycle No.	SW R1	Shorthand Description	Comments
		START	Operator pushes START button, connects the clock, and types in two numbers, 6 and 8 in sequence.
1.	A	(PCC)→MAR	PCC loads MAR to point at memory location 0000.
	B	(PCC)+1→PCC	PCC becomes 0001.
	C	((MAR))→IR	First instruction (location 0000) goes to IR.
	D	((Input 1))→ACC	Instruction performed; load first data word (input 1) into accumulator.
2.	A	(PCC)→MAR	MAR points at memory location 0001.
	B	(PCC)+1→PCC	PCC becomes 0010.
	C	((MAR))→IR	Second instruction (location 0001) goes to IR.
	D	((ACC))→210	Instruction performed; transfer data word from accumulator to location 210.
3.	A	(PCC)→MAR	MAR points at memory location 0010.
	B	(PCC)+1→PCC	PCC becomes 0011.
	C	((MAR))→IR	Third instruction (location 0010) goes to IR.
	D	((Input 2))→ACC	Instruction performed; load second data word (input 2) into accumulator.
4.	A	(PCC)→MAR	MAR points at memory location 0011.
	B	(PCC)+1→PCC	PCC becomes 0100.
	C	((MAR))→IR	Fourth instruction (location 0011) goes to IR.
	D	((ACC))+((210))→ACC	Instruction performed; add data word in accumulator to data word stored in memory location 210 and put total into accumulator (6 + 8 = 14).

(Continued)

TABLE 2.2. (*Continued*)

Master Cycle No.	SW R1	Shorthand Description	Comments
5.	A	(PCC)→MAR	MAR points at memory location 0100.
	B	(PCC)+1→PCC	PCC becomes 0101.
	C	((MAR))→IR	Fifth instruction (location 0100) goes to IR.
	D	((ACC))→output 9	Instruction performed; the data word (14) in the accumulator is outputted to the output unit.
6.	A	(PCC)→MAR	MAR points at memory location 0101.
	B	(PCC)+1→PCC	PCC becomes 0110.
	C	((MAR))→IR	Sixth instruction (location 0101) goes to IR.
	D	(HALT)	Instruction performed; disconnect clock to halt computer.

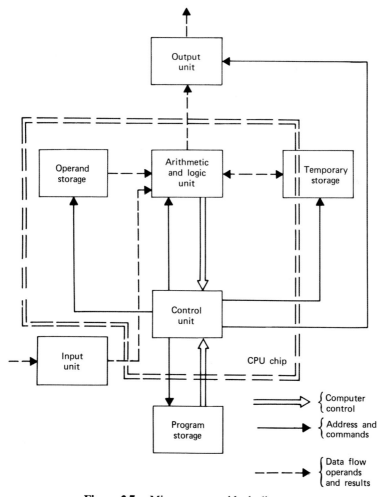

Figure 2.7. Microprocessor block diagram.

With the exception of providing the numbers to be added, no man–system relationship is required to perform the task of adding the two numbers together. The computer is operated totally on the fixed internal memory program, which is a permanent part of the system.

All microprocessors (as well as all larger digital computers) operate on the above procedure. The location of the individual units within individual integrated circuit chips may differ based on the device manufacturer, yet the general block diagram remains the same as shown in Figure 2.7. The detailed operation of each of the individual computer elements is examined in the subsequent chapters.

PROBLEMS

1. Describe the difference between an instruction step and a machine cycle step.
2. Show how the classical computer can combine the 10 basic arithmetic/logic functions to add three numbers together. Provide a complete set of minor cycle activities for this "program."
3. Demonstrate how the stack pointer keeps track of a program during the transfer-instruction step.
4. How are the program counter/stack pointer and program memory usually related?
5. What are the three primary uses of a computer's main memory?
6. What are the basic differences between a computer, minicomputer, and microprocessor?

CHAPTER 3

The Bipolar/MOS Technologies

The evolution of solid-state field-effect (MOS) electronics is not, as many people believe, a modern technology in the usual sense of the term. Studies of the field effect principle were undertaken and published as early as the 1930s by J. E. Lilienfield. Oskar Heil was awarded a British patent for a solid-state device in 1935. In 1948 at the Bell Telephone Laboratories in New Jersey, Bardeen and Brattain studied the effect of current modulation through point contacts on a germanium block when they discovered the point-contact transistor, a new phenomenon quite different from the field-effect principle.

This new concept changed the direction of semiconductor research and focused the full attention of the scientists on the development and application of the point-contact and bipolar transistor, with both of which the electronic industry is intimately familiar. Work on the field-effect principle virtually ceased.

It remained until the development of the silicon planar process in the early 1960s for the MOSFET to reawaken and emerge on the electronic scene as a viable semiconductor technology. This new technique of growing, etching, and regrowing an insulative oxide layer on top of a silicon substrate provided the required means of obtaining stable surfaces and controlling the physical geometry to the accuracy demanded for practical devices. However, problems in the control of charge migration rendered the practical FET device still far short of becoming a competitive technology with either bipolar techniques or vacuum tubes. In 1964, 30 years after the FET concept was discovered, the first simple MOSFET integrated circuits were produced.

The refinements of MOSFET manufacturing-processing techniques have since continued to provide rapid improvements in both MOS performance characteristics and yields. The recent developments in refined photolithographic equipment and techniques have resulted in the capability of producing complex photomasks with amazingly high resolution and dimensional stability necessary for the fabrication of large arrays, the backbone of the microprocessor technology.

Today the fields of MOSFET and bipolar technologies are both emerging as candidates for medium scale (MSI) and large-scale integration (LSI) application technologies. While the basic bipolar semiconductor (Fig. 3.1d) requires significantly greater square unit area than the smaller FET technology for its fabrication, its high speed amply justifies its added size. However, many applications, especially real-time servocontrol applications, do not require the high speeds possible with

17

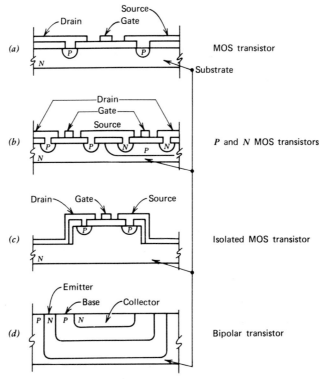

Figure 3.1. MOS/bipolar fabrication technology.

the bipolar components. It is in these applications that the MOSFET technology has emerged.

Because of the significantly smaller FET device size, when compared with its bipolar counterpart, the complexity of the FET circuit per unit area easily permits its entrance into the LSI field. Intimately coupled with this size reduction is a corresponding decrease in device power dissipation, permitting more devices to be crowded onto a single silicon chip than is feasible using the bipolar technique.

It is this combination of factors that has given a great impetus to the MOSFET over the bipolar technology within the field of the microprocessor. With the reduced expense of MOSFET technology, the desirability of dedicating the processor to a single task (as contrasted to multiplex time sharing of the minicomputer between many different and diverse tasks) is becoming more feasible.

THE BASIC MOS TECHNOLOGIES

The most widely used MOS transistor technology utilized at the present time is the metal-gate *P*-channel process. Cross sections of this type of device, called a MOS transistor, are shown in Figure 3.1*a–c*.

The basic structure is called the *substrate* and serves as a supporting body for the FET device. This thin slice, commonly referred to as the wafer during fabrication, is 8–10 mils thick and is lightly doped *N*-type silicon material.

The series element of the MOS transistor is composed of a source and drain (roughly equivalent to the collector and emitter of a bipolar transistor), which are heavily doped P-type regions, closely spaced, formed within the substrate by selective diffusion of a material that provides an excess of "holes" as its majority electrical carrier. These two regions are then separated by a thin deposit of metal, electrically isolated from the source, drain, and substrate by a thin layer of silicon dioxide (1000–1500 Å thick), which is called the *gate*. The isolation layer is often called the *gate oxide*. The gate serves as the control element for the MOS transistor creating a conductive or nonconductive channel between the source and drain regions, depending on the bias voltage on the gate.

The MOS transistor is completed by depositing metal contacts over the drain region and over the source region to provide for external connections.

OPERATION OF THE MOS TRANSISTOR

The operation of the MOS transistor is primarily a function of the modulation (or controlled variation) of the channel between the source and drain of the transistor. This modulation of the device is a function of gate bias. If a conductive channel exists with zero gate bias, the device is said to operate in the "depletion mode" (the saturated mode for bipolar transistor operation). If the device is normally nonconducting without gate bias, the channel being formed by applying a sufficiently negative voltage to the gate, the transistor is defined as operating in the "enhancement mode" (the cutoff mode for bipolar transistor operation). It can be seen that the enhancement mode is the lowest power mode of operation for the MOSFET since the absence of any gate bias (signal) forces the device into the "cutoff" condition and results in a zero-power dissipation for the device. This mode is extremely desirable for use in LSI digital circuits. In addition, this mode of operation is the easiest MOS transistor form to produce with the P-channel process.

The minimum gate voltage required to cause channel formation (current flow) is known as the *gate threshold voltage* and is a function of both the particular manufacturing process used and the lattice structure of the silicon ingot utilized for the substrate wafer. Voltages equal to or more negative than the threshold voltage level will cause the surface of the N-type silicon (substrate material) to invert to a quasi P-type material, hence giving rise to the term "P-channel."

The MOS transistor gate, being electrically isolated from the substrate, drain and source, has an input resistance on the order of 10^{15} Ω. This high input impedance yields an absolute isolation between the controlling circuit (gate drive circuit) and the control circuit (source/drain circuit). It also gives rise to the possibility of using essentially zero current drive circuits, better known as charge-transfer circuits.

The basic symbols used schematically to represent the MOS transistor are shown in Figure 3.2. It is understood that the body or substrate of the device is connected to the most positive potential for arrays fabricated with the P-channel process. A P-channel MOSFET device can also be utilized as an N-channel transistor by reversing the power supply leads and the polarity of external signals.

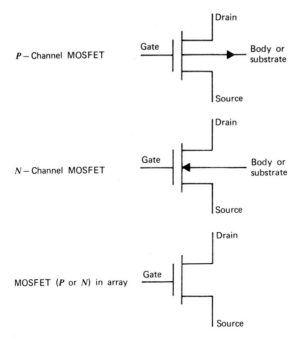

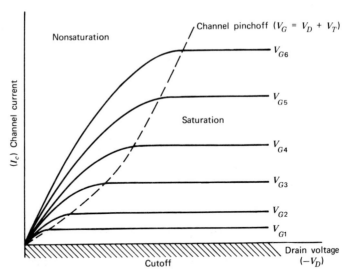

Figure 3.2. The MOS transistor.

ADVANTAGES OF MOS LSI

For the application engineer, the details of device fabrication ease and fabrication step reductions are of little consequence unless they are either more economical or show improved performance. These two basic advantages can further be broken into the following areas:

1. Economic factors
    a. Lower system cost
        i. System size

      ii. System power consumption
     iii. System layout
     iv. System fabrication
      v. System test
   b. Lower ownership cost
      i. Higher reliability
     ii. Improved maintenance

2. Performance factors
   a. Higher performance at same cost
   b. Greater reliability
   c. Controlled performance

Economic Factors

The economic factors of MOS LSI are potentially the most attractive of the advantages of economy and performance for this technology. The most obvious advantage is that of the lower system cost of a new system. For the new design, the following advantages must be considered.

System Size

The cost of any electronic system is a combination of the basic component costs, including the hardware and sheet-metal cabinet materials utilized in the system as well as the physical facility that must be made available for the system during normal operation. The earliest computers, built during World War II, were extremely limited in computational capability, yet required rooms full of equipment cabinetry to house the computer. As a result, the total cost of these early computers included the costs associated with the physical buildings required to house the unit. This physical facility cost was not only significant in the overall initial cost of the computer installation, but also contributed to the continuous operating costs of the installation.

With the advent of transistors, the reduction in the size of the computer system attributed to the use of transistors was soon offset by the desire for additional computational capability, especially in the area of memory capacity. However, with the advent of MOS LSI, technology has apparently overpowered the desire for additional capability, resulting in major reductions in the size of the overall computer systems and the corresponding decrease in the major costs of housing the system. The computational capability associated with the new programmable calculators is one such example.

In conjunction with the smaller system size associated with LSI computer systems is the reduction in the physical costs associated with the electronic cabinetry necessary to house the system. In many cases, especially with the hand-held calculators in wide use today, the size of the cabinet is ceasing to be a function of the magnitude of the internal electronics, but is rather a function of the man–machine input/output interface requirements such as the input keyboard and the output displays. It must be noted, however, that a computer of large physical size does have the advantage of maintaining inherent security (the potential theft of an electronic system becomes a significant problem as its physical size is diminished), while the significant decrease in physical size presents the added cost of providing protective security for the system, or insurance in the event of system loss.

A final result of the reduction in the system size is the decrease in individual component parts that must be employed in the system. This reduction in device count represents a large decrease in the absolute parts cost, which represents a major portion of the initial cost of the system, a cost that recurs each time a system is built. As time progresses, however, the initial cost of the components used in a complex computer system diminishes to a small fraction of the total system ownership cost, being overshadowed by the ongoing labor and material costs associated with the normal maintenance and operation of the system.

System Power Consumption

The second economic advantage of MOS LSI systems is the overall power consumption of the LSI computer system, which is lower than that in a system built out of comparable bipolar technology. Table 3.1 illustrates the comparison between modern MOS device technology power dissipation and comparable bipolar technologies. The MOS technology provides a 2.5–5.5 reduction of power over the bipolar technology. This power reduction results in a corresponding reduction in the size of the electronic power supplies necessary to power the system during normal operation. Such a decrease in power supply capacity and size provides not only a cost savings during the initial purchase of the equipment, but also a significant savings in the day-to-day operating power cost. An extreme example of this is seen in the modern microprocessor. The decrease in the overall electronic power requirement has permitted the introduction of the battery-powered calculator in contrast to the AC-powered calculator unit prevalent a few years ago. This trend toward AC/DC operation is steadily being introduced in larger, more powerful computers as component technology becomes available.

TABLE 3.1.   Comparison of MOS/Bipolar Power Consumption of a Single Gate

Process	Logic Family	Power (MW)
MOS	STD V_T—2 phase	4.0
MOS	Low V_T—2 phase	0.75
Bipolar	900 DTμL	8.5
Bipolar	54/74 TTL	10.0
Bipolar	54H/74H TTL	22.0
Bipolar	54L/74L TTL	1.0

System Layout

The introduction of MOS LSI device technology places the MOS device in the category of a black-box device, the majority of the circuit operations being performed within the LSI device with a corresponding minimization of external interface wiring connections. Thus, the printed circuit board becomes a support board for the LSI component with a minimum number of circuit interconnections between the individual LSI component chips. This is in contrast to the bipolar technology, which utilizes the printed circuit board as a wiring cable/interconnection board between the components in the circuit.

With the large-scale reduction in component count and intracomponent connections, the printed-circuit board layout is reduced to a simpler task for the design engineer, with a corresponding decrease in engineering costs (layout hours spent on the printed-circuit board design) as well as a reduction in potential circuit board redesigns.

It must be recognized that the black-box concept of LSI utilization also presents the possibility of reducing the need for costly multilayer printed-circuit board construction by permitting the use of the less costly two-sided printed circuit board. This is primarily due to the minimization of interconnection wires per functional block that is achieved with the MOS LSI technology as contrasted with the bipolar equivalent circuit.

System Fabrication

Once the design and layout of the printed circuit boards have been completed, the cost of system fabrication must be considered. Two main factors contribute to the cost of system fabrication: (a) cost of required material and (b) required number of construction man hours. Factor (a) has been primarily addressed under the evaluation of system size. As the MOS technology has ushered in the age of large-scale integrated technology, the overall cost and size of the system has shrunk even though the cost per component may have increased. Thus, the total overall cost of system-level components and material has decreased due to the vast decrease in the number of components required to build the system, a decrease in the number of printed-circuit boards required to hold the components, and a decrease in the size of the sheet-metal cabinet required to house the entire system.

Even more dramatic, however, is the decrease in the number of labor man hours required to fabricate the LSI system. This decrease in man-hour fabrication cost is most easily seen through the evaluation of the cost of a basic computer circuit as fabricated from the bipolar technology components and from the present LSI technology components. The single flip-flop design of the early 1960s is used for the present study. This circuit was composed of at least two semiconductor devices coupled with a minimum of four passive components (two resistors and two capacitors). Thus, six components had to be mounted for each flip-flop used in a memory/register application. For a 1024 × 8 bit shift register, 49,152 discrete components would be mounted on the printed-circuit board. It has been estimated that it costs an average of 5¢ to $1.00 per solder connection to install a component on a board. Using this estimate, a minimum total fabrication cost of over $4915 would be required to solder the components on a board for a 1024 × 8 bit discrete memory register, an exorbitant cost which would rule out the use of a discrete memory register of this size. This cost is in contrast to the present use of eight MSI devices (1024 × 1 bit MOS memory registers) costing approximately $8.00 to mount on the printed-circuit board (about 128 solder connections). This amounts to a savings of approximately $4900 in fabrication cost, not counting a significant decrease in the size and complexity of the printed-circuit module(s) over its discrete-component equivalent circuit. Comparable savings may be realized in every area of LSI MOS fabrication, from the memory plane to the input/output registers.

System Test

The savings to be realized during the system test phase of LSI MOS technology, as contrasted to a discrete bipolar system, is similar to the contrast found previously for the savings in fabrication costs between the two systems. If the same 1024 × 8 bit register illustration is used, there is a potential of 49,152 bipolar/passive devices that could fail in a discrete register system as compared with only eight for the LSI system. As a result, only eight LSI devices must be tested via automatic test equipment to isolate a system failure as contrasted with the complexity of testing the interaction of 49,152 devices to isolate the failed component.

The fault isolation of a failed device is virtually instantaneous with the MOS technology system, permitting an immediate replacement of the single failed component. In contrast, major trouble-shooting techniques and procedures are required to limit a system failure to a single component in the discrete circuit case. This is extremely expensive in terms of both time and dollars cost. Thus, while the LSI MOS components may be significantly more expensive from a single-component viewpoint, the ease of trouble shooting the failed LSI system and replacing the faulty device results in a total test time and replacement cost far below that required for the less expensive (per component) discrete bipolar system.

The above individual savings for the LSI MOS system are ultimately totaled and reflected as both initial and ongoing system cost savings to the purchaser. These costs are not, however, the only costs that the owner of a computer system must sustain. In addition to the initial cost, the routine maintenance of the system must be evaluated.

Greater Reliability

The primary operating cost for any system is directly related to the system's reliability. When a system fails, the overhead cost of the system to the owner remains intact while the return value is reduced to zero. In addition, the cost of service and repair must be borne by the owner (either by means of a service contract or direct service cost) in order to return the system to an operational state.

The LSI MOS reliability has proven to be far greater than its discrete bipolar counterpart. This increased reliability is due primarily to the reduction of interconnections between components that can fail, the decreased number of individual active and passive components (with their corresponding failure rates), and the simplicity of the MOS device construction as contrasted to the complexity of the bipolar device.

The weakest link in the reliability of any complex system is in the circuit interconnections within the system that are man-made. This weakness is primarily due to the fact that each connection is made by an individual who may have his attention diverted from the job during the period he is forming a circuit interconnection. (*Note:* This interconnect may be a solder connection, mechanical connection, wirewrap connection, or any other means of tying two electrical points together.) During this period of distraction, the quality of the interconnect is degraded and its potential for failure enhanced.

The vast majority of interconnections within the MOS system are machine-diffused interconnections in contrast to man-made ones. Instead of requiring more than 16,000 man-made interconnections to connect the bipolar discrete 1024 ×

8 register circuit, the use of the modern MOS chips decreases this number to approximately 108 man-made interconnections. This accounts for a 160:1 reduction in man-made interconnections with a corresponding increase in overall system reliability. An overall interconnection reduction of 100:1 is easily realizable for a moderately large MOS system including input/output peripheral equipment using today's LSI technology. Predictions of a 1000:1 reduction appear to be quite feasible in the near future.

Improved Maintenance

The greater ease of maintaining the LSI MOS system, as contrasted to the bipolar/discrete system using vastly greater parts count, has already been partially addressed under the ease of system testing. Let it suffice to comment that the logistics between maintaining an available supply of several dozen chip types for repairing a large computer installation, in contrast to maintaining a storehouse of discrete components for an equivalent discrete system, is immense both in dollar cost and facilities. In addition, the ease of replacing a single chip, as contrasted to hours or days of searching for a failed bipolar device, greatly overshadows any minor additional cost that the MOS system may incur.

Performance Factors

Higher Overall Performance

The final area of importance to the owner of a complex system is that of overall performance. While performance is a somewhat subjective gauge of a system, its economic consequences are not.

Performance ghosts within a large system can be extremely costly. Performance ghosts are those parameters that appear to be within operating tolerance when examined individually, but cause race problems, faulty triggering, and generally degraded system performance when combined within the overall system context.

The MOS LSI circuit, fabricated under rigid production controls that maintain and/or establish operating bias levels automatically, tends to be far more uniform in operating characteristics than multiple discrete circuits built using the same type of components. For this reason, fewer unpredictable performance problems arise in the LSI system than appear in the discrete system. In like fashion, aging in the LSI chip occurs uniformly over the entire chip and involves all components on the chip. This is not the case with a large discrete circuit involving many diverse types of components. Local heating often causes gross unbalance within the discrete circuit whereas the entire MOS LSI circuit is usually totally involved in the local heating condition, and the effects or temperature drift on the LSI circuits tend to cancel each other out with a result that the overall system performance remains relatively unchanged.

The Overall Trend

The trend of lower economic cost is on the side of the LSI MOS technology. The use of LSI MOS chips within the microprocessor field indicate a sizable cost reduction for the user in both initial and ownership costs. While these trends do not restrict future activities within the microprocessor field to the MOS technology, indications are that this technology will be the standard for determining system size and cost for years to come. Trade-offs involving speed and/or power consideration may lead the designer to the bipolar SSI or MSI technology, but such a conclusion will be made from the baseline of MOS technology and not vice versa.

PROBLEMS

1. What are the two primary reasons for using MOS technology for LSI chips?
2. What is the main reason for using bipolar technology at the present time?
3. Contrast the cost of a 64-bit memory system built in 1955 using discrete components to one available today. Use *NPN* transistors valued at 50¢ per device. It will cost 5¢ per solder connection to mount the parts, and the board will cost 50¢ per square inch to fabricate. The cost of a SSI 64 × 1 bit device will be approximately $1.00.
4. Using the 64-bit memory system constructed in question 3, estimate the time required to test the memory to determine if it is working. At $20.00 per hour, how much does it cost to test the discrete memory circuit?
5. If each part in the 64-bit memory system constructed in question 3 has a failure rate of one failure every 2000 hr, how many hours would this discrete memory be expected to operate before a failure would occur?

CHAPTER 4

The Memory

The system technologies utilized to construct electronic equation solving and/or decisionmaking equipment can be divided into two distinct categories: (a) real-time systems and (b) nonreal-time systems. The distinction between these two types is fundamentally associated with the time periods during which the input information is received, utilized, and outputted. If the flow of information through the system is *simultaneous* with the physical occurrence or generation of the information and the resultant output, information is a continuous flow of data based on the input information, although delayed by the propagation time through the system, the system is referred to as a *real-time* system. Most electronic control circuits and virtually all hand-held digital calculators fall into this category.

In contrast to the real-time system is the nonreal-time system. When the system uses data accumulated over an extended period of time, possibly modifies and subsequently stores unique or selected portions of that data, performs computations (either arithmetic or logical) on the data at a later time, and ultimately outputs the results of the computation on call from an external source without *any* dependence on the time of original data generation, the system is referred to as a "nonreal-time" system. Figure 4.1 details the comparison between the real- and nonreal-time computer data flow.

The prime example of the nonreal-time computer system is the modern large-scale digital computer. Data are collected by means of punched cards (or other techniques discussed later) and maintained in the punched card file (external memory) until the operator wishes to input the data into the computer. Once the operator inputs the punched cards into the computer, the computer stores this information in internal data registers or tape memories while it completes other computations. In a predetermined priority sequence, the computer ultimately performs the desired computation on the data. When the results of the calculations are complete, the computer may once again store the results in output registers or tape memories awaiting the operator's command to output the results in any one of many different formats. The overall flow of data is devoid of any relationship with real-time information flow, and thus the term "nonreal-time" system.

It is quite obvious that the nonreal-time system is uniquely tied to the large storage memory system. The size of the memory system that should be employed within a real-time system is more difficult to formulate. High-speed real-time sys-

Real-Time System	Nonreal-Time System
1. Input data directly to machine.	1. Input data placed into external memory.
2. Data flow simultaneous with input data.	2. Data input stored in internal memories.
3. Output offset from input by ripple flow only.	3. Data may be manipulated by computer before processing data.
4. Output immediately used by "programmer."	4. Computer processes data on "priority" basis.
5. All activities related to actual time.	5. Output data stored in external memory until called by programmer.
	6. No activity related to actual time.

Figure 4.1. Comparison of real and nonreal-time computer data flow.

tems with high ripple factors must of necessity minimize the utilization of storage elements due to the large time lost in these elements. For slower throughput (ripple factor) systems, the use of a memory system is more advantageous since more time can be devoted to the interrogation of the memory system as well as data transfer and manipulation. Thus, some real-time systems can utilize limited memory systems while many of the fastest real-time systems are forced to utilize circuit techniques.

THE MEMORY SYSTEMS

A memory system may be defined as any storage element for data. The written page is one form of a memory system utilized for remembering sequences of thought. The photographic slide or print is another form of commonly used memory system that is far more complex in nature than the written page due to the complexity of data being stored. In short, any technique that can be utilized to *remember* an event or data can be utilized as a memory system. However, the complexity of the memory system is directly proportional to the complexity of the data stored.

The memory system used in a digital computer is relatively simple, being required to *remember* only one of two states, a "zero" state (sometimes called the *low* or *false* state) and a "one" state (sometimes called the *high* or *true* state). The smallest digital memory element within the memory system is called a *cell* (or *bit*) and can be set in either a "zero" or a "one" state. (Future reference to these two states may be shown as a "0" or "1," respectively.) A group of these cells combined together form a digital "word" called a *byte*. Most digital computers utilize bytes that are combinations of 4, 8, 16, or 32 individual cells. This organization corresponds to the binary logic sequence as shown in Table 4.1.

The digital byte organization can be utilized for three basic applications: (a) numerical data values, (b) operation code designations (nonnumerical), and (c) memory, address locations (numerical). Since two of the three applications are numerical, the conversion between the decimal system and the binary system must be fully understood.

TABLE 4.1. Binary Word Organization

Number of Cells	Organization	Word Length	Range of Number
1	cell	1	0,1
2	byte/word	2	0–3
4	byte/word	4	0–15
8	byte/word	8	0–255
16	byte/word	16	0–65535
32	byte/word	32	0–4,294,967,295

Binary-to-decimal Conversion

The binary byte can be visualized as a series of digits (bits) beginning with the highest-value bit (called the most significant bit—MSB) and ending with the lowest-value bit (called the least significant bit—LSB). Figure 4.2 illustrates the typical byte configuration.

The decimal value of each individual bit is determined by its location n in the byte and by the absolute digital value of that bit. Thus, the value of an individual bit is determined by the equation:

$$a_n = |x| \cdot 2^n$$

where a_n = decimal value of bit n;

x = absolute value of bit as indicated by the bit value (0,1);

n = location of the bit beginning at the LSB = 0.

Referring to this equation, any bit having an absolute digital value of 0 ($x = 0$) has a numerical value of zero. In like fashion, any bit having an absolute digital value of 1 ($x = 1$) has a decimal value of 2^n. Figure 4.3 illustrates the values of 2^n for values of n between 0 and 64. It can be observed that the value of 2^{n+1} is twice that of the preceding value of 2^n.

The decimal value of a computer byte is equal to the summation of the values of the individual digital bits. Thus,

$$X = \sum_{n=0}^{n=\text{MSB}} |x_n| \cdot 2^n$$

where X = decimal value of byte

$n = 0,1,2,\cdots$

x = absolute value of bit

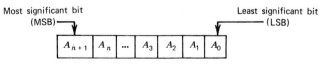

Figure 4.2. Typical byte configuration.

-n-	2^n
0	1
1	2
2	4
3	8
4	16
5	32
6	64
7	128
8	256
9	512
10	1024
11	2048
12	4096
13	8192
14	16384
15	32768
16	65536
17	131072
18	262144
19	524288
20	1048576
21	2097152
22	4194304
23	8388608
24	16777216
25	33554432
26	67108864
27	134217728
28	268435456
29	536870912
30	1073741824
31	2147483648
32	4294967296
33	8589934592
34	17179869184
35	34359738368
36	68719476736
37	137438953472
38	274877906944
39	549755813888
40	1099511627776
41	2199023255552
42	4398046511104
43	8796093022208
44	17592186044416
45	35184372088832
46	70368744177664
47	140737488355328
48	281474976710656
49	562949953421312
50	1125899906842624
51	2251799813685248
52	4503599627370496
53	9007199254740992
54	18014398509481984
55	36028797018963968
56	72057594037927936
57	144115188075855872
58	288230376151711744
59	576460752303423488
60	1152921504606846976
61	2305843009213693952
62	4611686018427387904
63	9223372036854775808
64	18446744073709551616

Figure 4.3. Positive powers of 2.

EXAMPLE.

To illustrate the conversion of a digital byte to a numerical decimal value, consider the 8-bit binary word.

$$a_7 \mid a_6 \mid a_5 \mid a_4 \mid a_3 \mid a_2 \mid a_1 \mid a_0$$
$$1 \mid 0 \mid 1 \mid 0 \mid 1 \mid 1 \mid 0 \mid 1$$

1 0 1 0 1 1 0 1

$$a_n = |x| \cdot 2^n$$

$$a_0 = |1| \cdot 2^0 = 1 \cdot 1 \quad = \quad 1$$
$$a_1 = |0| \cdot 2^1 = 0 \cdot 2 \quad = \quad 0$$
$$a_2 = |1| \cdot 2^2 = 1 \cdot 4 \quad = \quad 4$$
$$a_3 = |1| \cdot 2^3 = 1 \cdot 8 \quad = \quad 8$$
$$a_4 = |0| \cdot 2^4 = 0 \cdot 16 \quad = \quad 0$$
$$a_5 = |1| \cdot 2^5 = 1 \cdot 32 \quad = \quad 32$$
$$a_6 = |0| \cdot 2^6 = 0 \cdot 64 \quad = \quad 0$$
$$a_7 = |1| \cdot 2^7 = 1 \cdot 128 = 128$$
$$= 173$$

$$X = \sum_{a=0}^{a=7} a_n$$

Decimal-to-binary Conversion

The transformation from a decimal to binary number is the reverse of the procedure just followed to establish the decimal value of a binary number. If a decimal number is to be transferred into a binary number, the following steps should be followed:

1. Determine from Figure 4.3 the largest binary number (decimal value) contained in the decimal number being converted. This binary number is the most significant bit of the binary word. It is given a bit value of 1. All higher order binary bits have a bit value of 0.
2. Subtract the decimal value of the binary bit from the decimal number to obtain a decimal remainder.
3. Using the remainder, determine the largest binary number which has a decimal value equal to or smaller than the decimal remainder of two above. This binary bit is given a bit value of 1. The bit value of any binary bits between the binary bit determined in step 1 and this step receive a value of 0.
4. Subtract the decimal value of the binary bit determined in step 3 from the remainder determined in step 1. Use this remainder to repeat step 3 until there is no remainder. Set the bit value of any remaining lower order bits to 0.
5. The binary number is that word established by the bit values determined in steps 1–4.

EXAMPLE.

To illustrate the conversion of a decimal number to a binary number, consider the decimal number 196.

1. The largest binary number (decimal value) that can be contained in the number 196 is 2^7 (128). Thus, 2^7 has a bit value of 1, while 2^8 or higher level bits have a bit value of 0.

2. Subtracting the decimal equivalent of 2^7 (128) from the decimal number 196 yields a remainder of 68.

3. The largest binary bit number contained in 68 is 2^6 (64). Thus, the bit value of $2^6 = 1$. There are no binary bits between 2^6 and 2^7; therefore, no bit values are set to zero.

4. Subtracting the decimal equivalent of 2^6 (64) from the decimal remainder 68 yields 4.

5. The largest binary bit number contained in 4 is 2^2 (4). Thus, the bit value of $2^2 = 1$. The bit values for the binary numbers 2^3, 2^4, 2^5 are equal to zero.

6. Subtracting the decimal equivalent of 2^2 (4) from the decimal remainder 4 yields a remainder of 0. Thus, the bit value of 2^1 and 2^0 are set to zero.

7. The binary value of the decimal number is equal to the bit values thus calculated. Therefore,

$$196 = \begin{array}{c|c|c|c|c|c|c|c} 2^7 & 2^6 & 2^5 & 2^4 & 2^3 & 2^2 & 2^1 & 2^0 \\ \hline 1 & 1 & 0 & 0 & 0 & 1 & 0 & 0 \end{array}$$

The primary use of the binary byte is to designate the numerical value of a number. This value can be determined as indicated by the preceding discussion.

The second use of the binary byte is that of establishing control states for the control of the computer system through operation codes (see Chapter 2). Although the appearance of the digital word used as an operation code resembles a binary arithmetic data byte, its function is to control the operation of the computer rather than to perform arithmetic data computation (see Chapter 2). The use of the binary byte format as an operational code is fully discussed in Chapter 9 on machine-level programming.

Memory locations are generally addressed by use of a location word that again resembles a binary arithmetic data byte. The number of address locations that can be addressed by a single 8-bit byte word is 256 locations. If a two-byte memory address word is used, which is typical of most microprocessors, as many as 65,536 memory locations can be addressed. Each memory location addressed could contain a binary word having at least 8 bits.

The Intel 8008 utilizes a modified 2-byte memory address word, using 14 bits of data for address locations and the 2 MSBs for defining the operation to be performed. The 14-bit address word provides the capability of addressing a maximum of 16,348 unique memory locations.

The larger 16-bit processor machine frequently use a double 16-bit address word (32 bits), which serves to increase its potential memory address capability to approximately 4.3 (10^9) bytes of memory. A double 32-bit address word (64 bits) increases the available addressable memory to almost 10^{19} bytes. It is this large address word size that permits the general-purpose IBM type computers to simultaneously store and handle so many large programs.

TYPES OF MEMORY-STORAGE FORMAT

There are three unique and distinct types of memory system employed in the computer today. These memory formats are the: (a) sequentially accessed storage memory, (b) random-accessed storage memory (RAM), and (c) fixed data and control-storage memory.

Each of these systems utilize the identical word/byte format, storing each binary data byte in a unique addressable location. The primary difference in these three memory forms is the techniques used to access (reach the desired address location in order to be able to write data into the location or read data out of the location) the memory and/or whether the memory can be written into or accidentally destroyed during normal computer operation. These three basic forms of memory storage will be evaluated in Chapters 5, 6, and 7, respectively.

PROBLEMS

1. How large a number can be calculated by a 12-bit machine operating on a 12-bit byte?
2. If a machine uses an 8-bit word to address its memory, how many unique memory locations can be addressed?
3. Convert the following binary bytes to decimal values:

    a. 10110101   d. 11111111
    b. 01011011   e. 10111111
    c. 10111000   f. 11011001

4. Convert the following decimal values to binary bytes:

    a. 231   d. 137
    b. 124   e. 255
    c. 56   f. 13

5. If an 8-bit machine has an overflow bit (usually called a *flag bit*), can two numbers be added together (e.g., 245 and 221) without losing the total? (Assume the flag bit can be sensed during the "add" routine.) What is the largest number that can be calculated in an 8-bit machine with an overflow flag bit?

CHAPTER 5

The Sequentially
Accessed Storage Memory

The sequentially accessed storage memory is a memory in which the arithmetic or control data have been placed in a sequential line. When this type of memory is read, the sequential line flow is stopped, read incrementally by an appropriate form of reader, and then the line is incremented to the next data in the line and stopped again. This sequence is repeated until all data are read. The paper tape memory is the best known example of sequentially accessed storage. Each location corresponds to 5 or 8 bits of data comprising a single piece of information (either a digit or a letter). At time $T = 0$, the first digit of the group is read. The tape is then incremented under computer control one step and at $T = 0 + t$ ($t =$ time to increment the tape and prepare the read amplifiers for reading) the second digit of the group is read. This continues until all data have been entered into the machine.

Since these data are formatted as single pieces of information, they must be organized in a serial sequence, eliminating the possibility of reading the data at the end of the input tape (sequence) without passing through all of the intervening data locations. This gives rise to a large finite time delay in order to access or read any given piece of data from the sequential memory-storage system. Thus, the larger the sequentially accessed memory system, the longer the access time between the potential need for a given data word and the ability of the system to increment to the word location to provide the data.

Despite the great disadvantage of excessive access time delays in securing desired data, the sequentially accessed storage system is a commonly used memory system within computers today. The vast amount of data that can be stored, coupled with the nonvolatility of the data once stored, combine to offset the access time disadvantage. Finally, the cost per bit of this form of memory system is among the lowest for any type of memory system presently available, making it extremely attractive for large memory systems.

The main forms of sequentially accessed storage systems are paper tape storage, IBM type punched card, magnetic tape (open reel or cassette), disk memory (disk or floppy disk), bubble memory, and semiconductor shift-register memories (bipolar and MOS).

DATA FORMATS

Two data formats are utilized for storing data within the sequentially accessed storage. They are the bit-serial and bit-parallel, digit-serial formats. The difference between the two formats determines the form of input/output electronics that the systems require. With the bit-serial, digit-serial format (see Figure 5.1), the data word is presented one bit at a time to the output electronics, requiring four clock periods to read a single 4-bit word and eight clock periods to read a single 8-bit word.

With the bit-parallel, digit-serial format, the data word is presented to the output electronics with all bits in parallel (see Figure 5.2). This requires four (or eight) output amplifiers (compared to the need for only one output amplifier for the bit-serial, digit-serial format) to be used simultaneously, permitting the full data word to be transmitted from memory in only one clock time.

The great speed advantage of the bit-parallel, digit-serial format over the bit-serial, digit-serial format is obvious. Although there is a great increase in economic cost due to the amount of electronics required to gain this increase in speed, the

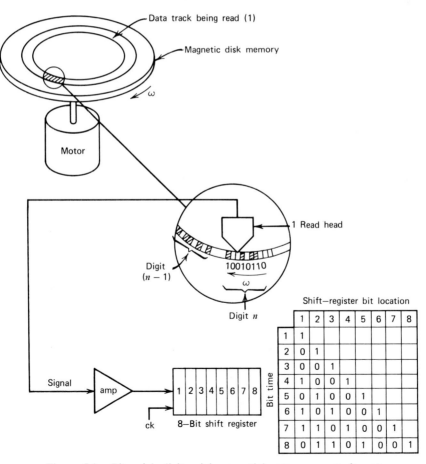

Figure 5.1.   Bit-serial, digit-serial sequential-access memory format.

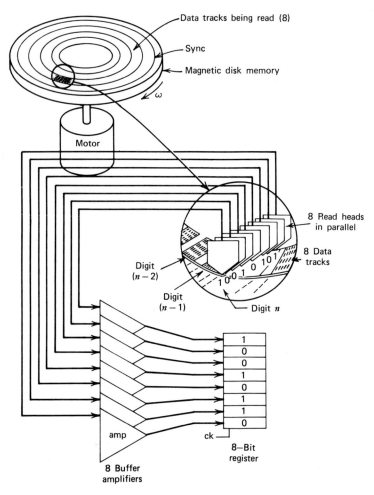

Figure 5.2. Bit-parallel, digit-serial sequential-access memory format.

lower overall cost of the mass memory system may well offset this increased expenditure.

With the exception of the bubble and shift-register memories, the mass memories are all electromechanical in nature and operate on a similar principle. As previously stated, the primary attraction of these memory systems is the vast amount of memory locations that can be provided at a low cost. The overpowering disadvantage beyond the long access time encountered with the sequential system is the problem of mechanical adjustment, critical mechanical tolerances, and periodic maintenance ever present in any electromechanical system.

THE SEMICONDUCTOR SHIFT REGISTER

With the advent of the semiconductor shift register, an exit from the smaller electromechanical temporary storage systems has been made to the self-contained semi-

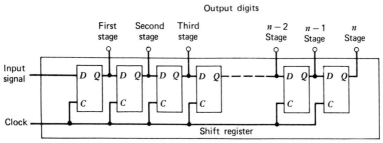

Figure 5.3. Flip-flop shift register.

conductor-type memory. The shift register is a combination of flip-flop cells placed in series and sequenced by the application of a common clock signal (see Figure 5.3). The input-data bit sets the state of flip-flop #1 to the desired state on the application of the appropriate clocking signal (t_0). At the next clock signal (t_1) the data bit is transferred from flip-flop #1 to flip-flop #2, flip-flop #1 returning to the "0" state in the absence of a "1" level at the input. At the clock signal t_2, the data bit is transferred from flip-flop #2 to flip-flop #3, and so on until the bit has been transferred the entire length of the shift register. The state of the final flip-flop is monitored and its output state read out at the appropriate time. Unless additional external circuitry is used, the data information is destroyed when it is transferred out of the final shift-register flip-flop. There may be, however, provisions made to provide a feedback circuit from the final flip-flop output to the input flip-flop of the shift register creating a circulating memory system (see Fig. 5.4). An alternative to employing immediate feedback from the output to input flip-flop cells of the shift register is to series two or more shift registers together in a daisy-chain fashion to provide almost an endless length shift register. All shift registers in the daisy chain must be controlled by the same clock signal. By appropriate feedback, the multiple shift-register memory can also be transformed into a recirculating memory system.

The basic single shift register format is bit-series, digit-series. It can, however, be easily transformed into a bit-parallel–digit-serial format by paralleling M registers together and clocking all the registers simultaneously. Thus, a $(N \times A) \times M$ parallel shift-register memory system can be created as shown in Figure 5.5.

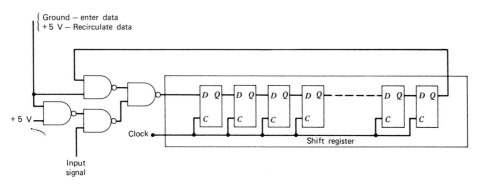

Figure 5.4. Recirculating shift-register circuitry.

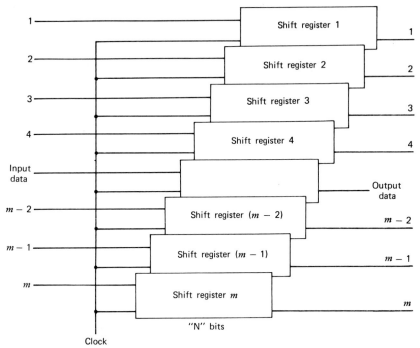

Figure 5.5. Parallel shift register $(m \times N)$.

For small memory systems, the $(N \times A) \times M$ bit-parallel–digit-serial semiconductor memory has many advantages in terms of reliability, initial cost, and future expandability. In addition, the capability of intermixing long and short loops within the same memory system provides a great flexibility over the disk memory, which must complete a full revolution before returning to a given data location (assuming no increase in electronics) and as a result has a constant-length loop.

It is obvious, however, that the semiconductor shift register memory system will remain a poor second choice to the magnetic tape or disk-memory systems for large-scale memory systems for the foreseeable future. To meet the mass memory market, Gerald Luecke, Jack Mize, and William Carr of Texas Instruments estimate that the shift register would have to provide bit densities of a million bits per square inch at a cost of tens of millicents per bit and provide an access time of less than 8.5 msec. to any data point in order to significantly impact the present demand for the large mass memory. Although the semiconductor shift register is short of being competitive within the mass memory market, it remains very attractive and competitive for the small to medium memory market. It seems extremely probable that the bubble memory will emerge from the laboratory within the relatively near future to fill this gap.

PROBLEMS

1. Describe and illustrate the operation of a bit-serial, digit-parallel tape recorder.
2. Indicate which form of shift register you would use, and why, in the following applications:

    a. single-wire data communication;

    b. typewriter storage register;

    c. temporary storage memory for a microcomputer.

3. Name four types of sequentially accessed memories that can be used.

4. Design a recirculating memory for holding 2048 bytes of memory (1 byte = 8 bits long) and describe how it works. Assume that there is a 64 × 1 bit shift register available.

5. Describe the major reason for using the sequentially accessed memory. Name four magnetic types in popular use today.

CHAPTER 6

The Random-Access Storage Memory

The second memory format of importance is the random access memory (RAM), gaining its name from its ability to read out of or write into any byte location within its memory system instantaneously. The main distinction between the RAM and the sequentially accessed memory is the term *any* location within the memory. Thus, instead of waiting for a *sequence* time period to elapse between the request to address a memory location and the physical addressing of the desired location, the random access addressing of the desired location is instantaneous. The accessing of any location is totally "random" in nature. This ability of instantaneous accessing is credited with increasing the overall computer performance speed to those associated with the modern computer.

The random access storage memory has proven practical, utilizing only three forms of technology: (a) magnetic cores, (b) plated wire, and (c) semiconductors (both bipolar and MOS). A general view of organization of an RAM system is shown in Figure 6.1.

It can be seen from the comparison of Figure 6.1 with Figures 5.1 and 5.2 that the RAM involves significantly more physical electronics to perform its function than the sequentially accessed memory previously evaluated. However, the immediate recall of any cell location(s) within the memory matrix is often such an important function in the overall operation and value of the memory system that the additional cost of the required support electronics (both in component cost and increased area required for mounting and interconnecting the components) becomes insignificant.

CELL ADDRESSING

To select a designated cell, two coordinate select signals are required. One signal selects the x coordinate of the memory matrix, while the second signal selects the y coordinate. When these two signals are applied to the matrix simultaneously, the cell(s) is ready to either receive information from the outside (referred to as "writing" information into memory) or supply information to the outside (referred to as "reading" information from memory). The selection of the read or write command is under external control and is normally provided by a single input terminal signal.

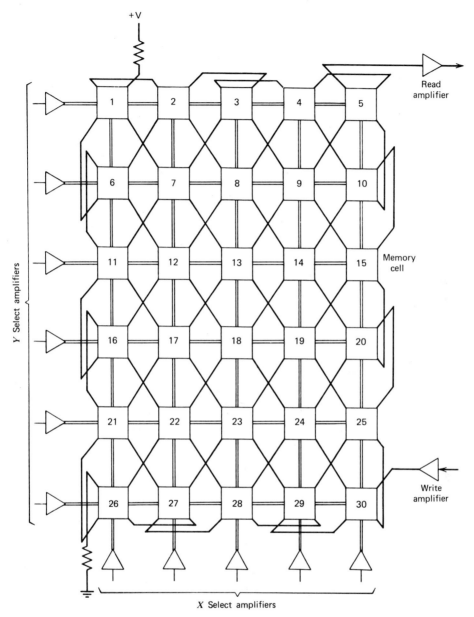

Figure 6.1. Random-access memory—general organization.

    The selection of the x and y coordinates of the memory is normally based on an n-bit binary selection command providing the location of the column and row of the cell being addressed. Thus, for a 4×4 memory matrix such as shown in Figure 6.2, the row (y) address is contained in the first 2-bit $(a_1 \mid a_0)$ binary word (four possible states), while the column (x) address is contained in the second 2-bit $(a'_1 \mid a'_0)$ binary word (four possible states). These two address words may be combined into a single address byte of four bits $(a_1 \mid a_0 \mid a'_1 \mid a'_0)$ the first two bits

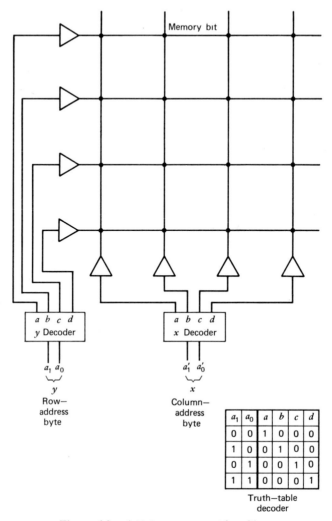

Figure 6.2. 4×4 memory matrix address.

a_1	a_0	a	b	c	d
0	0	1	0	0	0
1	0	0	1	0	0
0	1	0	0	1	0
1	1	0	0	0	1

Truth—table
decoder

defining the row (y) coordinate and the last two bits defining the column (x) coordinate. Thus, the minimum size binary byte required to provide the address for addressing a 16-bit memory is 4 bits long.

The memory matrix may be expanded from a $(n \times m) \times 1$ bit memory to a $(n \times m) \times d$ memory by paralleling d matrix memories, where d corresponds to the number of bits in the desired memory storage word. This parallel byte matrix configuration for a $4 \times 4 \times 4$ matrix is illustrated in Figure 6.3. The desired byte location within the matrix is addressed by the identical address word as used to address a single bit location, the d parallel address lines (x,y) of the byte being simultaneously addressed in common. Each parallel matrix read/write amplifier will be unique, reading out of or writing into the individual d bits of the parallel byte. In order to support this increased byte length, d sets of unique electronics will be required, one set for each digital bit.

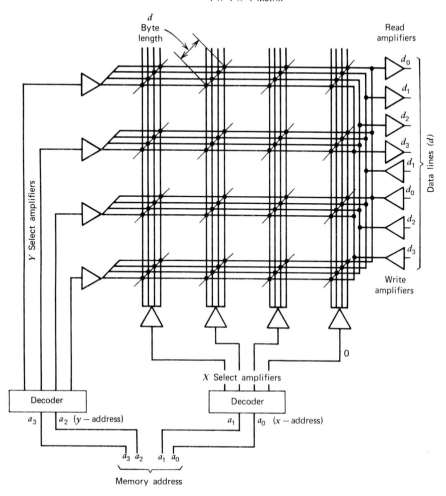

Figure 6.3. Parallel byte matrix.

It can be extrapolated that the maximum size of the RAM memory matrix is a direct function of the size of the computer binary address word. An address word of 8 bits can address a matrix of either 256 rows × 1 column, 256 columns by 1 row or any combination of rows and columns making up a total of 256 cell locations. For example, a matrix of 64 rows by 4 columns can also be addressed by an 8-bit address word.

The expanded memory formed by the use of parallel "word" bits (bytes) is addressed simultaneously as the "base matrix" bits are addressed. Thus, if eight 256 × 1 bit matrix devices are placed in parallel and the x,y address words are made common to each, the 8-bit address word controls 2048 bits of data (i.e., 256 8-bit words or 256 bytes) rather than merely 256 bits of data. It is obvious that provisions must be made in this system to simultaneously read out of or write into the 8 bits selected. Figure 6.4 illustrates this type of memory organization.

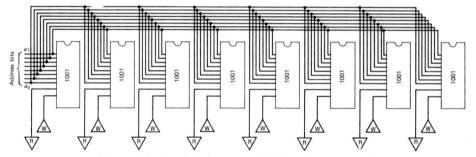

Figure 6.4. Semiconductor parallel byte memory matrix.

Figure 6.5 illustrates a third type of memory organization commonly referred to as a *three-dimensional matrix memory*. This memory differs from the previous expanded-memory system only in that the parallel bits (words) are physically placed in the third dimension leaving addressing capability on the x–y coordinates. This form of memory organization has advantages in magnetic core arrays but is normally not utilized in the semiconductor memory arrays.

It should be noted that the decoding of the memory address, and the subsequent x and y line selection, can utilize a significant portion of the total electronics associated with the memory system. An 8-bit decoder necessary to transform an 8-bit binary code address to one of 256 lines requires over 37 separate DM 7223 (or equivalent) 1-line to 8-line demultiplexers, or 17 separate DM 54154 (or equivalent) 1-line to 16-line demultiplexers. This complexity corresponds to over 425 individual amplifier/gates. In discrete form, this memory decoding circuit would be inordinately large and costly.

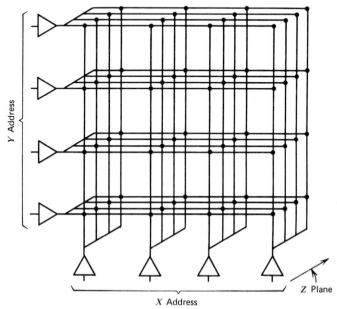

Figure 6.5. Three-dimensional RAM matrix.

THE MAGNETIC CORE MEMORY

The first practical RAM was constructed utilizing magnetic cores as the memory element. The square-loop magnetic core has unique physical properties that enable its use as a memory device. Figure 6.6 illustrates the ideal square hysteresis loop required to create the magnetic core memory elements. The initial write cycle forces current through the core in either a direction to force the core from A to BCD (defined as a "1"), or from A to FGH (defined as a "0"). Once set, when the core is read, it is forced to the zero state from the state that it presently holds. Thus, assuming it is currently in the "1" state, during the read cycle the core will be electrically pulsed negatively by the read current, forcing the core's magnetic state from D thru EFG to end at H. During this transition, the change in magnetic flux experienced causes a large current to be sensed by the sense winding and a "1" level to be read from the core. If, however, the core was initially in the zero state (H), the core will be forced from H to J by the read current pulse causing a

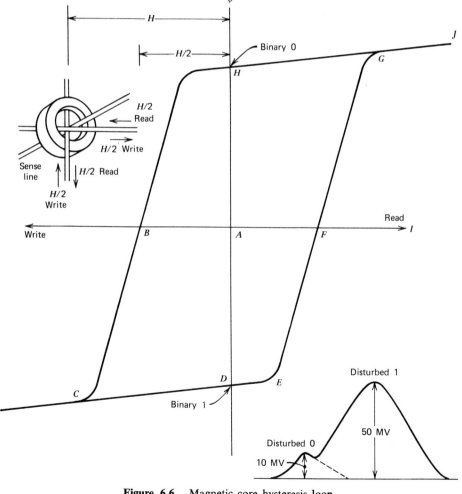

Figure 6.6. Magnetic core hysteresis loop.

small magnetic flux change to be sensed and a zero ("0") level to be read out. In both cases, the read current forces the core to the zero state.

The major disadvantage of the core memory is that the *reading* process is volatile (destroys the existing data in memory). That is, the reading of the state of the core always leaves the core in the zero state regardless of the core's original state. In order to make the core memory appear to be nonvolatile, the state of the core before reading must be refreshed (read back into the core) each time the core is interrogated. This added requirement for immediate rewriting of data into the core after each read cycle requires the memory core cycle time to be extended beyond the time to "read" the core as well as requiring significantly more electronics to perform this rewrite function. Consequently, the magnetic core memory is a relatively slow cycle time memory system.

The primary advantage of the large core memory stack lies in its low cost for core material and its high reliability. However, the large amount of support electronics required to supply the read and write core switching currents as well as to refresh the memory after reading has forced the overall core memory costs to be extremely high for small memory systems. As the core stack is increased in size, the increase in the cost of the support electronics is minimal. Thus, for large-size memories, the core stack memory system is attractive.

The reliability of the core memory system follows the identical pattern as the cost of the system. For small memory systems, the support electronics necessary to operate the stack create an undue reliability loss on the overall stack. However, as the size of the stack is increased, the quantity of support electronics does not significantly increase, but the overall reliability of the stack dramatically increases. An example of a typical core memory stack is shown in Figure 6.7. Although many techniques have been innovated to make core memories both faster and smaller, the introduction of new higher-speed semiconductor memories was predicted as the end to the magnetic core memory system. Yet, with the introduction of still better and smaller memory cores, as well as the three-dimensional memory array, the magnetic core memory still maintains a viable position in the field of large RAM arrays.

Today the magnetic core memory addresses the core to be written into or read out of by means of the X,Y matrix selection only. The X and Y core currents, when summed together, are just large enough to cause the selected core to switch. Some manufacturers provide an increase in this margin of safety of the write current magnitude by placing a negative current through the unselected lines to further buck out the influence of the row and column line currents and thus ensure that an unselected core cannot inadvertently switch. In addition, one sense-coil wire is threaded through all the cores of the stack and connected to a single output read amplifier, thus minimizing the total read electronics required to refresh the core memory. Reading and writing is presently accomplished by reversing the directions of the currents in the X,Y selection amplifiers. These changes in technique have reduced the overall support electronics required by the core stack to a manageable level for moderate-size memory systems.

SEMICONDUCTOR RANDOM-ACCESS MEMORIES

With the introduction of the semiconductor RAM the small-size computer became a reality in terms of capability and price. The initial semiconductor memories were

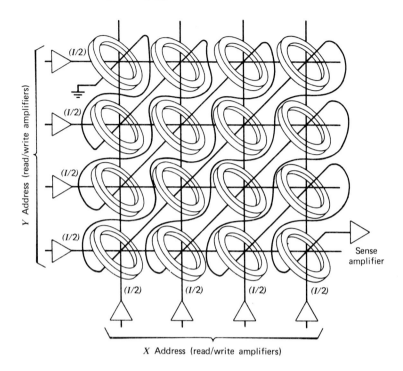

X Address (read/write amplifiers)

Figure 6.7. Core memory stack. *Note:* For core to be set to 1 or to be read (reset to 0) the current through the core must be equal to $I = (I/2 + I/2)$.

introduced as bipolar devices utilizing internal flip-flop memory circuits and having the necessary address-decoding circuits within the device package itself. This revolutionary small-size semiconductor memory device greatly simplified the printed circuit board layout for small and large memory arrays (sometimes called memory *planes* when referring to the semiconductor memory system). Figure 6.8 illustrates the internal circuit for a 256-bit RAM memory of the 74200 type, while Figure 6.9 illustrates the external printed circuit board connections necessary to operate this memory chip.

The Bipolar Memory Cell

The typical bipolar memory cell is schematically shown in Figure 6.10*a,b* and illustrates the simplicity of the bipolar RAM memory. Figure 6.10*a* illustrates the dual emitter flip-flop cell arrangement. It should be noted that one of the multiple emitters of each transistor is attached to the word line, while the remaining emitters are connected to one of the bit lines (a noncommon line). The community of flip-flops is thus interconnected by the word and bit lines to form the larger array. To read an existing state from this cell, the sense amplifier corresponding to the bit line or cell column is activated (set to read) while the word line is raised in voltage. As the word line is increased in voltage (the word line normally carries the emitter current from all the cells in a row), the current through the emitter of the "on" transistor is transferred from the word-line emitter to the bit-line emitter. The bit-line sense amplifier then senses the resultant voltage due to the transfer current

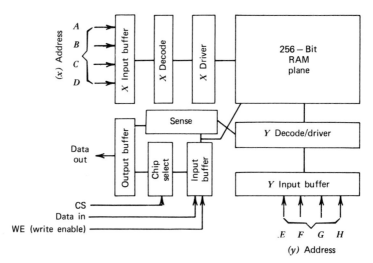

Figure 6.8. 256-Bit RAM Memory 74200.

as either a one (1) or a zero (0). No information, it should be noted, is destroyed during this read cycle.

To write into the cell a similar selection process takes place. The "write" amplifier connected to the bit lines is offset, causing one of the two emitters to be more negative than the others. Simultaneously, the voltage of the word line of the selected cell is raised above the voltage of the bit line, permitting the negative bit line to control the state of the cell. Since the voltage on the nonselected cell word line remains lower than the most negative bit line, there is no change of state for these cells. This cell technology is used by manufacturers such as Intel, Intersil, Computer Microtechnology, and Raytheon.

The alternate cell configuration illustrated in Figure 6.10*b* utilizes a pair of Schottky diodes to minimize cell size. To read from this cell, the word line is lowered in voltage (as contrasted to raising the word-line voltage as in Figure 6.10*a* technique) and the voltage offset is sensed on the selected bit line by means of a current unbalance through the Schottky diode. Writing is accomplished in much the same way. The word line is first lowered and the selected write amplifier feeds a large current into one of the bit lines (denoting a 1 or a 0) through the

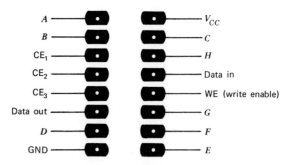

Figure 6.9. 74200 Printed-circuit board connections.

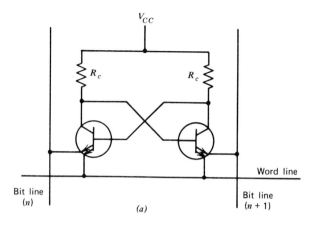

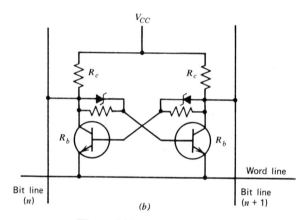

Figure 6.10. Bipolar RAM cell.

Schottky diode, which turns on the transistor to which it is connected and loads down the opposing transistor's collector impedance, thus forcing it to turn off while the selected device is driven on harder. This technique is used by Intel in its 256-bit 3102 memory.

The outstanding advantage of the bipolar semiconductor RAM is the high speed that the bipolar transistor array is capable of reaching. The outstanding disadvantage, however, is the large cell area that must be provided for the bipolar devices and, due to their complex structure, the low yield in production and resultant high overall cost. Since the basic cost per chip is high, the cost of the bipolar semiconductor memory is inherently high.

The MOS Memory Cell

An alternative to the complex bipolar structure is the simpler MOS structure device. Figure 6.11 illustrates the schematic of a MOS memory cell. The two sides of the flip-flop, Q_1/Q_2 and Q_3/Q_4, are controlled by the gating devices Q_5 and Q_6, respectively. When the cell is being read, the word line turns on both Q_5 and Q_6,

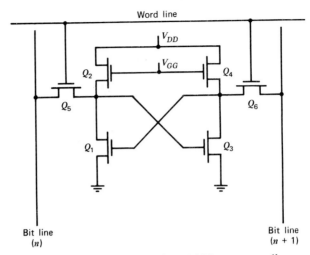

Figure 6.11. Six-transistor MOS memory cell.

transferring the state of the cell to the bit lines to be sensed by the selected read amplifier. During the write time period, the word line again turns on the control gating devices Q_5 and Q_6 and forces the cell into the desired state by setting the proper voltage at the selected bit lines. This form of MOS six-transistor memory cell is utilized in virtually all existing 256-bit MOS memories on the market today.

All of the above bipolar and MOS devices are "static devices." That is, they utilize a bistable flip-flop storage element for the memory cell, which, when set, remains set until intentionally changed. This type of static storage cell has two gross disadvantages:

1. It continuously dissipates energy, which automatically limits the number of cells (bits) that can be placed into a given package size.
2. It occupies a large amount of silicon area, which also limits the number of bits that can economically be put onto a single monolithic package.

To minimize the power dissipation for the MOS-type transistor, the V_{GG} supply line may be clocked on and off. Referring to Figure 6.11, whenever the V_{GG} line is held negative, Q_2 and Q_4 will become active load elements and full power will be dissipated by the cell. When, however, V_{GG} is held positive, the load elements Q_2 and Q_4 are turned off (open-circuited), and no further power is dissipated. The correct memory information will be maintained in the cell by charge storage only. That is, one side of the flip-flop was negative prior to the turnoff (clocking) of the V_{GG} supply. This side will maintain its negative potential by storing a negative charge on the gate capacitance of the other flip-flop transistor in the memory cell. When the Q_2/Q_4 power elements are again activated, the cell will again return to its desired state.

If the impedance of the MOS device were infinite, the stored negative charge on the gate capacitance would be maintained indefinitely. However, in parallel with the transistor parasitic gate capacitance is the junction impedance associated with the cross-coupled transistor device. This junction impedance produces a leakage

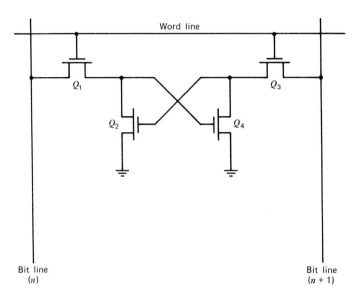

Figure 6.12. Four-transistor MOS memory cell.

path for the stored charge to discharge through the reverse-biased transistor junction. Thus, the stored charge in the memory cell has a definite time duration and requires periodic refreshing in order to retain the integrity of its memory state.

The required periodic refreshing of the MOS memory cell is provided by periodically clocking the V_{GG} bias level to its negative voltage which activates the load elements Q_2/Q_4 and replenishes the lost negative charge through the load elements. Such a memory device is called a *dynamic charge storage memory,* or *dynamic memory* for short, in contrast to the static memory device, which does not require any refreshing of the memory state.

The second and perhaps most serious disadvantage of the MOS transistor type of memory element is not resolved by changing from a static to dynamic clocking technique. This disadvantage is the relatively small packing density possible due to the use of the large size six-transistor memory cell. A higher-density memory can be achieved, however, by changing from the six-transistor flip-flop to a smaller four-transistor cell shown in Figure 6.12. This cell is basically a flip-flop without load devices.

With the four-transistor cell, Q_1 and Q_3 take the place of both the load devices and the word-enable control transistors. To read the state of the cell, the bit-line current is sensed by the appropriate read amplifier. If Q_2 is on, the bit line associated with Q_2 will carry current (Q_1 also being in the "on" state) while the bit line associated with Q_3 and Q_4 will carry no current. To write into this memory cell, the sense lines are placed into their proper relation and then transmit the desired data into the cell by enabling the word line. To refresh the cell, the bit lines are both placed at a common negative potential and the word line is turned on.

An even greater reduction in the size of the memory cell can be achieved beyond the four-transistor memory cell. The complex flip-flop circuit is not necessary to store charge in a capacitor and to sense the presence or absence of that charge.

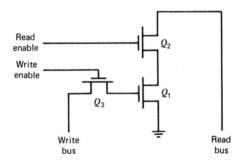

Figure 6.13. Three-transistor MOS memory cell.

For a single P-channel MOS device, any negative charge on its gate will cause the device to turn "on." Lack of a negative charge on the gate of the device will cause it to turn "off," thus creating a "0" and "1" memory storage element. Unless the gate charge is changed intentionally, the MOS device will retain its memory and thus fulfill its requirement as a memory cell. Figure 6.13 shows the configuration of a three-transistor charge-storage cell.

The transistor Q_1 becomes the charge-storage device for the cell, while transistor Q_2 forms the read-out path from Q_1 to the read bus when activated by the read-enable signal. Transistor Q_3 provides the write path to the charge-storage node when activated by the write-enable signal as well as the periodic refreshing path for the memory data. In the refreshment mode the data are read and then re-written into the cell by an existing on-chip refresh amplifier.

Figure 6.14 illustrates the typical memory organization for a three-transistor cell array. It can be seen that each column of memory cells has its own refresh amplifier. Activating any one of the read-enable signals automatically refreshes all of the memory cells in that row by rewriting each cell individually. Refreshing the entire memory thus takes refreshing n rows since all devices in that row are simultaneously refreshed. This configuration is used by the Intel 1103 (1024 $\times$ 1) dynamic RAM device. Other variations of the three-transistor memory cells have been produced, but they are merely modifications of the basic circuit form shown above.

The RAM Memory Planes

The most popular form of RAM memory chip to be used in building the memory systems is the $n \times 1$ bit chip ($n = 256, 512, 1024, 2048$). While there are several chips of different configurations than the $n \times 1$ chip configuration, these have not been accepted as enthusiastically as might be expected.

To construct a useful computer memory plane, a word length (byte) of greater than 1 bit must be formed. The microprocessor memory plane is generally constructed with a word length equal to the computational byte used in the ALU. Thus, for the byte used in the 8-bit microprocessor, a memory word length of 8 bits would be used. A notable exception to this rule is the 4-bit microprocessor (e.g., Intel 4004 or 4040) that is a 4-bit machine but uses an 8-bit memory sys-

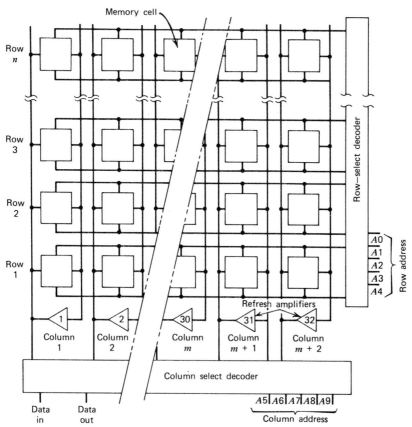

Figure 6.14. Typical MOS memory organization.

tem and the 'bit-slice" machines that can be pieced together to form virtually any size (word size) machine desired. For the "bit-slice" machine, the memory system word size will be equal to the combined machine word length.

The desired word length for the memory system is constructed by mounting m memory chips in parallel, m being the number of bits in the word length (usually $m = 8,12,16,24,32,64$). For an 8-bit microprocessor, eight "$n \times 1$" memory chips would be mounted in parallel to build a "$n \times 8$" memory plane. Each data line, (one data line going to each individual chip) would remain separate from the others and would be directly connected to the appropriate data bus line ($d_{m-1} \cdots d_0$). The address and read/write lines for the eight parallel chips would be common to permit all of the chips to be addressed simultaneously. Figure 6.15 illustrates the construction of a "$4n \times 8$" memory plane, where n equals the number of bits contained in a single chip.

Memory-cell Addressing

The addressing of the memory plane is generally performed by means of a double-length word. For the 8-bit microprocessor system, the use of a 16-bit address

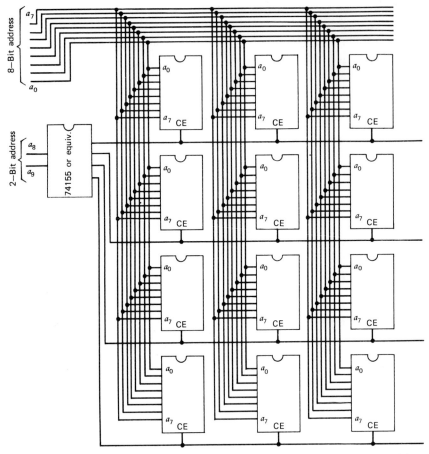

Figure 6.15. Basic memory plane (1024×8) using 1101 (256×1) RAM chips.

yields 65,536 unique addressable memory bytes. The two address bytes are designated as the low-order address word and high-order address word (see Figure 6.16).

Consider the two 8-bit address bytes as equivalent to a 16-bit counter with the count entering at the a_0 bit. By entering 65,536 count pulses, the entire count range will be incremented from 0 through 65,535. The next count pulse will reset the counter back to 0. However, during this incrementing process, the lower-order count (8 bits) will be incremented from 0 to 255 for each change in the higher-order (8-bit) count, the higher-order count being incremented by 1 each time the lower-order address returns to zero. On this basis, each higher-order count (address) contains the full 256 locations of the lower-order count (address). In the 16-bit address, the lower-order address (a_0–a_7) is called the *word (byte) address,* while the higher-order address (a_8–a_{15}) is called the *page address*. For the 8-bit memory system, a page contains 256 bytes (2^n bytes, where $n = 8$) of word address while there can be as many as 256 pages (2^n pages, where $n = 8$).

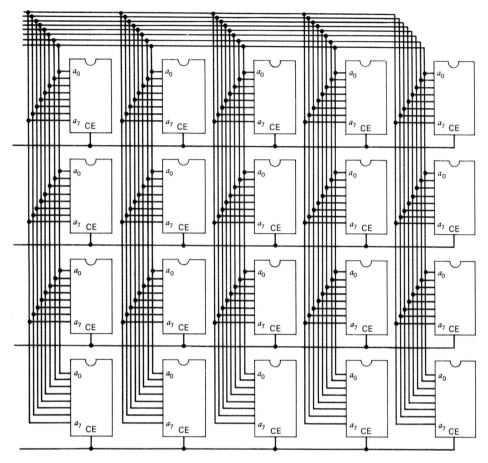

Figure 6.15. (continued)

The word-address lines are common to all pages of the memory system, while each page is controlled by its own page address (see Figure 6.17). In reality, the distinction between word address and page address is more useful in the software programming of memory than in the memory board layout since a memory chip containing more than 256×1 bits is addressed by more than the lower order (8-bit) address. A 1024×1 bit chip must, for example, be addressed by a 10-bit

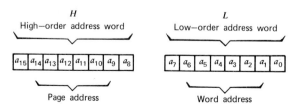

Figure 6.16. Two-byte address word.

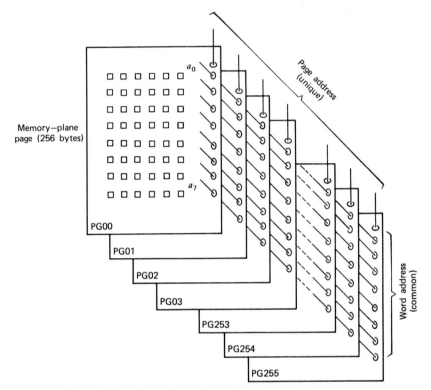

Figure 6.17. Comparison of word and page address structure.

address code, the chip containing a portion of four memory pages. Figure 6.18 illustrates the differences in addressing techniques used in addressing a 2048-byte memory plane when using a 256 × 1 bit chip configuration and a 1024 × 1 bit chip configuration. Note especially the use of the chip-enable (CE) input to select the desired page(s) to be addressed.

The full 65,536-bit memory plane can be constructed in a similar manner using the decoder scheme shown in Figure 6.18. Figure 6.19 illustrates the page decoder for a full addressable memory using the 256 × 1, 1024 × 1, 2048 × 1, 8192 × 1 and 16,348 × 1 chip configurations. The latter two chip configurations are still on the drawing board, but will soon be available in the marketplace.

An alternate technique of decoding the page address is the use of printed circuit jumpers installed into the circuit board. These jumpers connect the appropriate address lines to the desired decoder lines on the printed-circuit board. This technique avoids the costly transmission of multiple "enable-chip" signals throughout the system and reduces the addressing cables to the primary 8-bit word and 8-bit page-address lines. Figure 6.20 illustrates this form of P.C. board decoder for a 2048-byte memory card fabricated using the 1024 × 1 bit chip configuration.

Table 6.1 lists a few of the more popular RAM chips currently available. It should be noted that at least one major new memory chip enters the market each month, many of which are uniquely tied to the microprocessor system for which

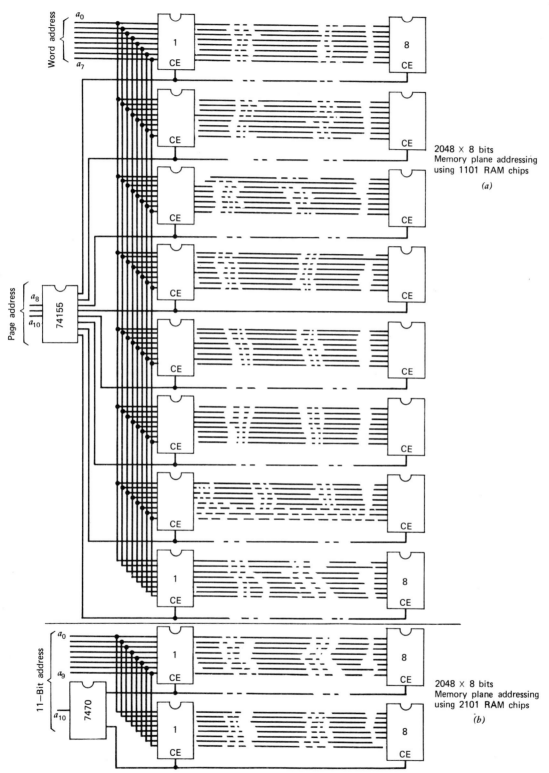

Figure 6.18. Comparison of addressing complexity of 2048-byte memory plane using 1101 (256 × 1) and 2102 (1024 × 1) RAM chips.

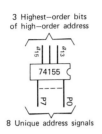

3 Highest—order bits
of high—order address

a_{15} a_{13}

74155

P7   P0

8 Unique address signals

1-74155 (3:8 demultiplexer) chip required to address 65K memory using 8K RAM memory chip

1-74155 (2:4 demultiplexer) chip required to address 65K memory using 16K memory chip (use only a_{14} and a_{15} address bits, and P0-P7 address signals)

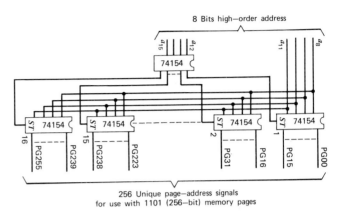

8 Bits high—order address

a_{15} a_{12} a_{11} a_{8}

74154

ST 16   74154   ST 15   74154   ST 2   74154   ST 1   74154

PG255 PG239 PG238 PG223   PG31 PG16 PG15 PG00

256 Unique page—address signals
for use with 1101 (256—bit) memory pages

17-74154 (4:16 demultiplexer) chips required to address 256 pages of memory. (65K memory) using 1101 (256 × 1) RAM chips

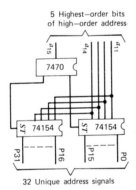

5 Highest—order bits
of high—order address

a_{15} a_{14} a_{11}

7470

ST   74154   ST   74154

P31   P16   P15   P0

32 Unique address signals

1-7470 (Flip-Flop) chip and 2-74154 (4:16 demultiplexer) chips required to address 65K memory using 2048-bit RAM memory chip

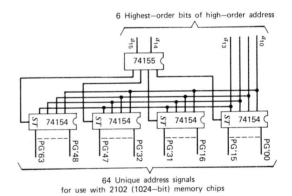

6 Highest—order bits of high—order address

a_{15} a_{14} a_{13} a_{10}

74155

ST   74154   ST   74154   ST   74154   ST   74154

PG'63 PG'48 PG'47 PG'32 PG'31 PG'16 PG'15 PG'00

64 Unique address signals
for use with 2102 (1024—bit) memory chips

1-74155 (2:4 demultiplexer) chip and 4-74154 (4:16 demultiplexer) chips required to address 65K memory using 2102 (1024 × 1) RAM memory

Figure 6.19. Comparison of decoding techniques required to address full 65K bytes of memory when using various-size RAM chips.

they were designed. Table 6.1 is restricted to a listing of those memory chips that are of a general nature and can easily be utilized with multiple systems.

The listing in Table 6.1 does not include RAM chips designed primarily for unique system usage [i.e., Motorola MCM6605L (4096 × 1 RAM) and National Semiconductor Corp. pace chips].

TABLE 6.1. Popular RAM Chips

Part Number	Technology	Configuration
3101	Bipolar	16 × 4
3101A	Bipolar	16 × 4
3106/7	Bipolar	256 × 1
3106A/7A	Bipolar	256 × 1
1101A	Static MOS *P* channel	256 × 1
2102	Static MOS *N* channel	1024 × 1
1103	Dynamic MOS *P* channel	1024 × 1
2107	Dynamic MOS *N* channel	4096 × 1

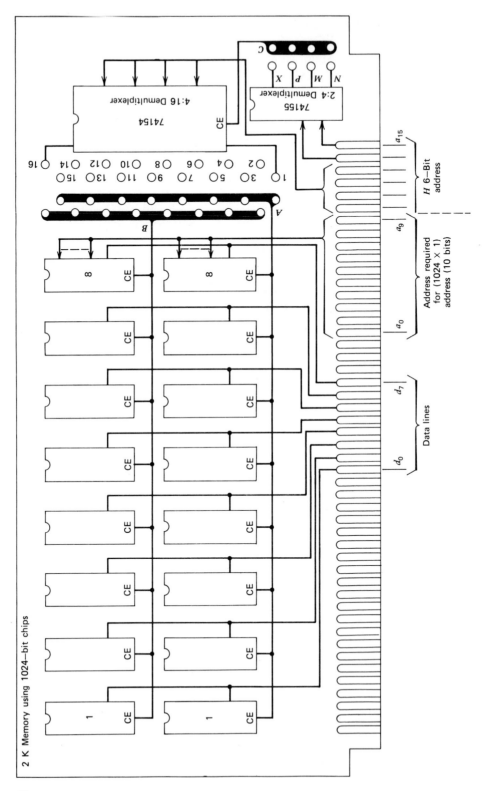

2 K Memory using 1024—bit chips

60

Figure 6.20 — "Universal" 2048 byte memory board

A and B Interconnects

C Interconnects	A/B	1	2	3	4	5	6	7	8	9	10	11	12	13	14	15	16
N ← C	A	0–3		8–11		16–19		24–27		32–35		40–43		48–51		56–59	
	B		4–7		12–15		20–23		28–31		36–39		44–47		52–55		60–63
M ← C	A	64–67		72–75		80–83		88–91		96–99		104–107		112–115		120–123	
	B		68–71		76–79		84–87		92–95		100–103		108–111		116–119		124–127
P ← C	A	128–131		136–139		144–147		152–155		160–163		168–171		176–179		184–187	
	B		132–135		140–143		148–151		156–159		164–167		172–175		180–183		188–191
X ← C	A	192–195		200–203		208–211		216–219		224–227		232–235		240–243		248–251	
	B		196–199		204–207		212–215		220–223		228–231		236–239		244–247		252–255

Figure 6.20. "Universal" 2048 byte memory board using 1024×1 chips. Memory page truth table. EXAMPLE: To establish memory for pp. 144–147, connect P to C and 5 to A.

PROBLEMS

1. Describe the operation of a computer read cycle when using a RAM.
2. Describe the operation of a three-transistor MOS memory cell during the *write* cycle, *refresh* cycle, and *read* cycle.
3. Design a 1024 × 8 bit memory plane using 256 × 1 bit memory chips. If each chip (MM1101) has 16 pins and occupies an area of one (1) square inch, and assuming a cost of 30¢ per square inch of circuit board, $1.00 per component (RAM and decoder) and 5¢ per solder connection, calculate the cost of this memory system.
4. Design a 1024 × 8 bit memory plane using a MM2102 (1024 × 1 bit) memory chip. Using the same costs as indicated in problem 3, except for the cost of the MM2102 ($4.00), calculate the cost of this memory system.
5. Design a complete memory system for use with a 16-bit (two 8-bit words) address capability using the MM2102 (1024 × 1) RAM memory chip. Determine the physical size of your design and the approximate cost using the costs indicated in problem 4.

CHAPTER 7

Static Data and Control Memories

The memory systems examined thus far have been the RAM type. In many of the larger computer installations the RAM memories, either rotational or core memory systems, are also used for the maintenance of static data and/or computer-control programs. In these larger systems the use of nonvolatile (data in memory are not lost when the power to the computer is removed) RAM systems permit permanent data storage at a minimum cost per bit and can serve to more efficiently utilize the overall general memory system.

If, however, the semiconductor RAM memory systems are employed, the stored material (data and program control logic) **is** lost whenever the power to the computer is turned off. Thus, a new type of semiconductor memory system is called for, a semiconductor system that is nonvolatile.

In addition, if the stored information is permanent (mathematical reference tables, character generation codes, etc.) or a permanent computer program is desired, it is conceivable that the capability of accidentally writing into memory may be very undesirable. Thus, a second feature of the new type of memory is established, a read-only capability.

The above two conditions may be combined to produce a second semiconductor form of memory system that is designated READ-ONLY MEMORY or ROM. The ROM system can be developed by either the modification of an existing semiconductor RAM system or utilizing specially designed semiconductor ROM memory components.

MODIFIED RAM SYSTEM

If a ROM memory is to be constructed from an existing RAM system, it must first be made nonvolatile with respect to the main power-supply source. Secondly, the capability to write into the RAM memory must be disabled.

The first requirement, nonvolatility, calls for the development of a secondary passive power supply and possibly a refresh clocking system if the dynamic RAM system is used in lieu of a static RAM system. Figure 7.1 illustrates one secondary power-supply technique that has been utilized in several computer systems. Although the secondary power voltage level is approximately 0.7 V lower than the primary

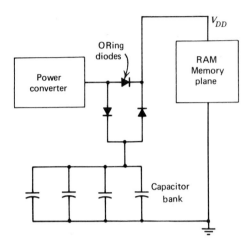

Figure 7.1. Modified RAM system (ROM).

power voltage level supply due to the "oring" diode, it is hardly worth expanding any additional design effort on increasing the voltage amplitude of this back-up supply, since its voltage is high enough to guarantee the memory's integrity, assuming the primary power is not left off too long.

The primary concern in the secondary power system is power capacity. For the secondary power system used to protect against momentary power failure back up, the capacity of the system can be attained by the use of multiple large capacitors arranged in parallel. Such a system could maintain the integrity of the ROM for several hours without any loss of data. However, should there be a need for a secondary power back-up system extending into weeks or months, the capacity must be extended by the use of batteries rather than capacitors. This solution will increase the overall size, weight, and complexity of the ROM and must be seriously evaluated before such an approach is undertaken.

The second requirement of the modified RAM system, the disabling of the write capability of the RAM, presents a simpler design problem and solution than the secondary power-level requirement. Using a discrete switch as shown in Figure 7.2 illustrates the simplest solution to this requirement. During the normal operation of the ROM, the switch on the front panel will be set to ground (zero volts), forcing the memory into a "read only" configuration. If there arises a requirement to modify the memory, the front panel switch may be turned to the +5 V state, which will permit the memory to be written into. Once changed, the panel switch will again be returned to ground to reestablish the ROM configuration of the memory system. There should be provided, however, a mechanical cover over the panel switch or other form of circuit interlock to prevent accidental switching "on" of the write signal and the accidental loss of the ROM data.

It may be concluded that, while the RAM memory system can be modified to form a ROM memory system, the cost of the overall support electronics can become quite excessive and the design considerations (weight and size) overpowering. An alternative solution to ROM memory design should be considered.

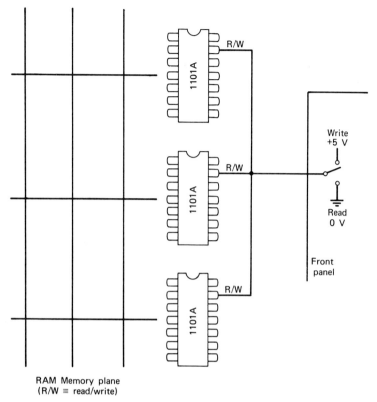

RAM Memory plane
(R/W ≡ read/write)

Figure 7.2. Modified RAM write interlock.

SEMICONDUCTOR ROM SYSTEMS

A more acceptable solution to the fabrication of a ROM memory system is the use of a specially designed semiconductor component called a ROM. The two design requirements, nonvolatility and nondestructability (unable to be written over), are both satisfied by this unique component design.

Four types of ROMs are presently available to the circuit designer: (a) the mask-programmable ROM, (b) the field-programmable ROM (PROM), (c) the electrically programmable ROM (PROM), and (d) the erasable/electrically programmable read-only memory (EROM).

The Mask-programmable ROM

The read-only memory (ROM) is composed of a permanent group of addressable digital words (control commands or data sets) that are combinations of binary 1 and 0 values. Based on the required nonvolatility, nondestructability of the ROM, a simple ROM can be constructed using the diode technique illustrated in Figure 7.3. The selection of the desired word is established by electrically forcing

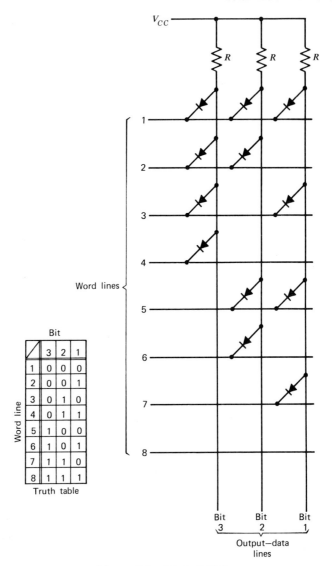

Figure 7.3. Diode ROM.

the selected "word" line low (to a negative voltage or ground). A diode located at the matrix cross point and joining the bit line to the word line will conduct, causing the voltage at the output of the bit line to go toward ground resulting in a "0" bit being outputted. If a diode is not located at the cross point (or the diode lead is broken), zero current will flow through the bias resistor (R) and a corresponding "1" bit will be outputted at the output data line.

Figure 7.4 shows a matrix similar to the diode matrix constructed by the use of MOS transistors instead of discrete diodes. By the selection of the appropriate "word" line, the desired MOS transistor(s) is biased into the conduction mode. The selection of the appropriate y select line provides the connection of the matrix transistors (series devices) to the output sense lines. When the desired transistor

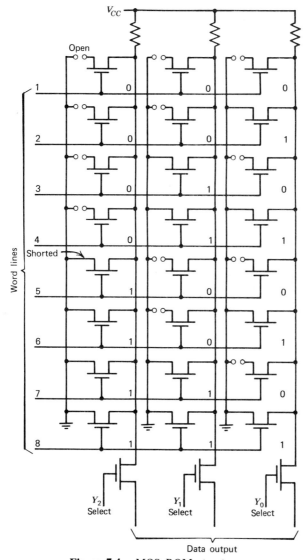

Figure 7.4. MOS ROM structure.

word line is grounded, the selected sense line will be shorted to ground if the metalization on the drain of the MOS device exists, yielding a "0" output. Alternatively, the selected sense line is isolated from ground (a high impedance is provided to ground) if the metalization to the drain of the MOS device is missing and a "1" is sensed at the output.

Both the ROM diode or MOS transistor matrix can be fabricated by the use of "mask-programming" techniques, although this technique is usually reserved exclusively for the MOS transistor arrays. In the "mask programming" technique, the basic MOS array is prefabricated on the chip (dice when still in the die form) without any metalization plated to the drain, gate, or source areas. Once the desired word pattern is received from the customer, the metalization pattern mask is prepared to correspond to the desired logic pattern, the device metalization being deposited (connected) if a "0" is being formed and the device left unconnected if a "1" is desired. It can easily be appreciated that the drain and gate metalization interconnections are also unnecessary if the device is not being used to form a particular bit. Thus, only those devices that must operate (provide a "0" level) are interconnected by metalization.

Since the metalization step is an integral part of the final fabrication step of the integrated circuit, this type of ROM can only be fabricated and programmed by a device manufacturer. As a consequence, if the mask-programmable ROM is programmed incorrectly, it must be discarded. Since it is a unique part (i.e., only one customer has a need for a ROM programmed with the special byte sequence), there is a large initial cost associated with the special handling of this type of device. Thus, for small numbers of ROM devices, the cost for each ROM device becomes prohibitive.

In most cases, the mask programmed ROM will require a masking charge (the cost to initially layout, check, and produce the metalization mask associated with a new program sequence) of $500 or more and a minimum purchase of 10–25 devices if required. This initial cost becomes insignificant if several thousand devices are to be produced using the same metalization mask, but is exorbitant if only a few devices are needed. While other techniques of fabricating mask-programmable ROMs are available, the overall approach and economic considerations are identical.

The Field-programmable ROM

In contrast to the mask-programmable ROM, which can be programmed only during the final stage of the device fabrication, a field-programmable ROM can be programmed by anyone who owns a special programming console. This form of ROM device is extremely important for use in small-quantity prototype production runs or system-development activities.

Two types of device fall within the field programmable ROM family. The first is the bipolar diode array as shown in Figure 7.5 and suggested in Figure 7.3. The second is a bipolar transistor configuration similar to the mask-programmable MOS transistor array except using bipolar fabrication techniques.

Both forms of the field-programmable ROM are constructed using the technique of deposited "fusible links" formed in the device metalization in series with the base/emitter junction of the bipolar transistor or the P/N junction of the diode. All of the diodes composing the ROM matrix are connected from the appropriate bit lines to the word lines by means of the fusible links. The fusible link is composed of a constricted portion of the interconnect metalization, which will fail (burn open) on the application of a pulse of energy, typically a pulse of 50–100 μA for a period of 2 μsec. This energy pulse is sufficient to melt (blow) the metalization at the fusible link and open the diode circuit. After the link is opened, the drive pulse is continued for a short period of time in order to keep the metalization from solidifying in a shorted link condition and reconnecting the blown fusible link.

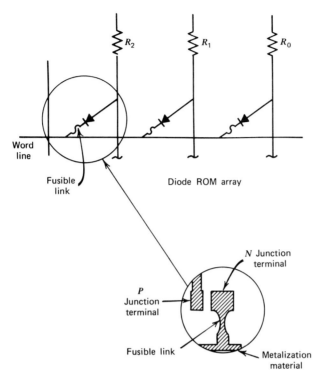

Figure 7.5. Bipolar diode programmable ROM.

Like the mask-programmable ROM, an error in the programming of the field-programmable ROM cannot generally be corrected. If a mistake is made in the ROM by the opening of a link by mistake, the ROM becomes worthless. To correct the mistake usually requires a new ROM device to be programmed.

The field-programmable bipolar ROM programming circuit suggested by Intel for use in the programming of their 3601 (256 by 4 bit) bipolar ROM chip is shown in Figure 7.6. This programming circuit is only one of many which are recommended by the various manufacturers. Each ROM device type may require a significantly different ROM programming-circuit configuration since the energy required to "blow" open a fusible link interconnect is a direct function of both the fabrication techniques and device technology used by the individual device manufacturer. As a result, the programming circuit that is valid for programming one ROM device may not be acceptable for programming another ROM device, even though the ROM devices are electrically interchangeable in the memory system.

The physical programming of the PROM by means of a standard programming circuit/console is extremely laborious and time consuming. As a result, many programming errors can occur, resulting in lost time and material costs. Unless an automatic programmer is utilized, each subsequent PROM will in turn be individually programmed like the first. Such a tedious programming procedure is obviously acceptable only for the initial ROM prototyping activity or for a very short preproduction run. The economics of using the mask-programmable ROM for long production ROM runs is obvious.

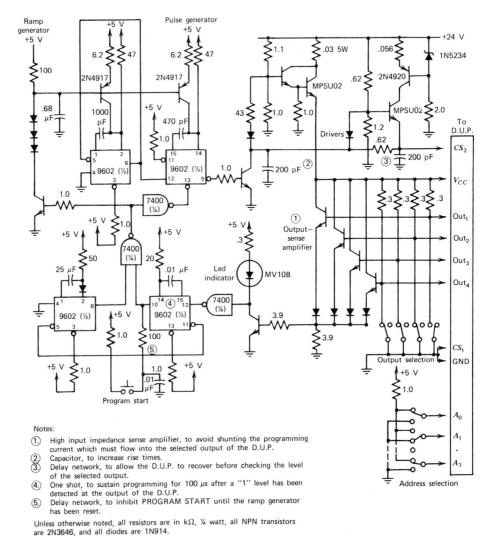

Figure 7.6. Intel 3601 programmer. Copyright Intel Corporation. Reprinted by permission.

The Erasable–Programmable ROM (EROM)

A variation to the field programmable ROM is a field-programmable ROM that can be *erased* and reprogrammed like the "fusible link" ROM. This ROM device is called an EROM (erasable read-only memory). This erase feature is extremely valuable during the development of experimental ROM programs since an error in the ROM program does not destroy the device, but merely requires the device to be erased and reprogrammed.

The EROM can be fabricated in two distinct types of MOS arrays or as a glassy semiconductor-type ROM. The latter ROM devices are usually packaged in a dual in-line package with a transparent lid.

The first type of MOS EROM device is fabricated like the typical MOS array except that a thin layer of silicon nitride is deposited between the metal gate region and the oxide insulator region of the device. This silicon nitride layer has the property of retaining an electrical charge (either positive or negative, depending on the type of material used to fabricate the MOS transistor array) when the transistor gate is subjected to an external high-voltage programming pulse. This programming pulse is several times the magnitude of any voltage level normally seen by the memory device. In the absence of any additional programming signals or sustaining power, the charge in the silicon nitride layer will remain virtually indefinitely. The presence of the charge will create the opposite state for the bit (transistor) being programmed. That is, a programmed one (1) pulse at the gate of the transistor being programmed will cause the normally nonconducting transistor to turn *on,* resulting in a zero (0) output bit. The device that has not been programmed will retain zero charge in the silicon nitride layer and remain *off,* resulting in a one (1) output bit.

The predominant advantage of this type of EROM construction is in the capability of the device to be reprogrammed (modified) by means of applying a reverse-polarity voltage pulse to the transistor gate to be corrected in order to remove the trapped charge in the silicon nitride layer. This returns the junction to its original nonconducting state. Since the programming (or reprogramming) voltage pulse is applied to the gate of the transistor bit being programmed, the input programming pulse is isolated from the programmed bit by the high-impedance oxide insulation material. As a result of this isolation, the programming/reprogramming of this type of ROM can be accomplished without removing the EROM from the memory circuit board. This permits the EROM to be an *in computer* programmable ROM. The EROM can also be reprogrammed by the individual byte, modifying a single word bit pattern at a time. This type of EROM can be fabricated using either *N*- or *P*-type material resulting in a nonprogrammed output bit being either a "1" or a "0" (the nonprogrammed output equal to "1" was described).

The second type of MOS EROM is composed of an array of *P*-MOS floating-gate transistors, all transistors being initially in the nonconductive (1) state. The application of a high voltage between the source and drain of the device removes the positive carriers from the floating gate, the isolated negative carriers remaining trapped on the gate. The resulting negative charge creates a conducting channel from the source to drain in the device. In this respect the device resembles the EROM previously described.

This second type of EROM, however, cannot be erased by the application of a reverse voltage pulse between the source and drain of the device, since a conductive path has been established in the transistor (due to its programmed state) and the device would be destroyed by the application of the reverse voltage pulse. Rather, the application of ultraviolet light through a quartz lid on the device is used to discharge the gate potential and return all of the device transistors to the nonconduction (1) state. This reprogramming is conducted on the device external to the computer and returns all the device bits to their initial "1" state (nonconducting transistor state). Room light, sunlight, or normal fluorescent lights have no measureable effect on the stored data, even after prolonged exposure. The application of ultraviolet light for 10–20 min. will, however, completely reset the programmed bits to their original state.

The third type of EROM, the bipolar EROM, utilizes a glassy semiconductor material that can be electrically altered from a high-impedance state to a low-impedance state by means of the application of controlled voltage pulses. The results obtained using this type of ROM device compare favorably with the above MOS EROMs. Yet, this type of EROM has not found a large acceptance, as compared with the other two EROM types, within the electronic community. As a result, this form of EROM will most likely become obsolete.

THE PROGRAMMED LOGIC ARRAY (PLA)

The programmed logic array (PLA) represents a modification of the programmable ROM. That is, the PLA is a programmable diode gate-array device that can be programmed as combinations of "OR" gates or "AND" gates. By the utilization of simple logic combinations, most forms of diode gate logic may be realized. The PLA array differs from the ROM device only in its overall organization, being available as both mask- and field-programmable PLA devices.

An example of the field-programmable PLA is shown in Figure 7.7. The array contains diodes with rows of common cathode connections and columns of common anode connections, each diode containing a fusible link in series with it. Figure 7.8 illustrates the technique typically utilized to form a series of "AND"

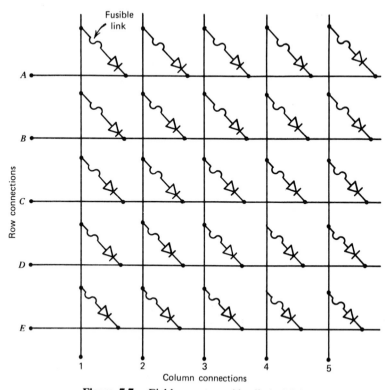

Figure 7.7. Field-programmable diode PLA.

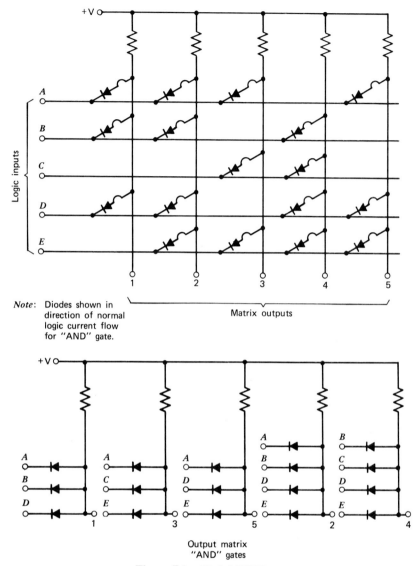

Note: Diodes shown in direction of normal logic current flow for "AND" gate.

Matrix outputs

Output matrix "AND" gates

Figure 7.8. PLA "AND" gate.

gates. The matrix is modified by the addition of load resistors connected to the common anode columns forming the traditional diode "AND" gate.

If the matrix is modified by adding the load resistors to the common cathode column, the diode array is transformed into an "OR" gate diode matrix as shown in Figure 7.9. A combination of "and" and "or" gates may be obtained by the intermixing of input and output rows and columns as shown in Figure 7.10. The "AND" gates are formed by the fifth row becoming the "ORED" output terminal. The logic diagram is shown in Figure 7.11.

An alternative use of the PLA may be achieved by combining it with a demultiplying chip such as the MM54181 (4–16 demultiplexer chip) to transform the

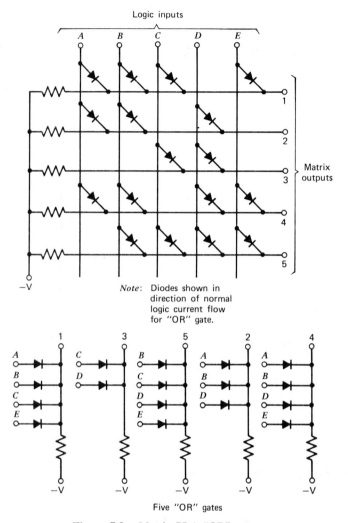

Figure 7.9. Matrix PLA "OR" gates.

PLA into an inexpensive ROM. The columns of the PLA become the output bit terminals of the ROM while the rows become the selected word terminals. When a selected matrix diode remains intact, the corresponding bit level will be shunted to ground through the diode and a "0" bit outputted. However, when the diode's fusible link is opened, the V_{cc} level "1" is directly outputted indicating a "1" bit.

While the PLA cannot compete with the PROM for large memory systems, it does merit consideration for application in small memory systems in which the capacity of the PROM is not fully utilized and the cost of a ROM is not justified. The PLA may also be useful as a permanent external logic control circuit within the microprocessor's interface circuits. This is the application for which the PLA was originally designed. By applying the fixed PLA logic control circuit prior to the input registers of the microprocessor, the apparent speed of the overall system can be significantly enhanced. This added speed is primarily due to the logic func-

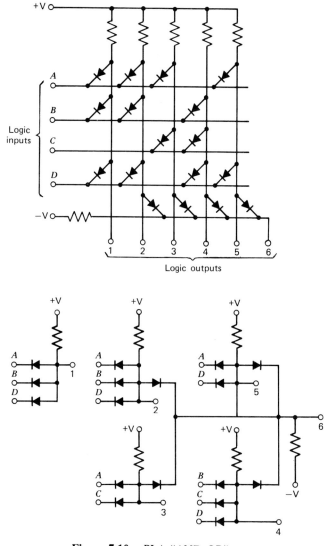

Figure 7.10. PLA "AND–OR" gates.

tions of the PLA being performed in real time while the results are operated on by the microprocessor at its machine cycle rate.

The PLA has been shown to be a very versatile device, but has failed to capture the imagination of the application engineer at large. Part of this failure is due to the fabrication difficulty experienced by many device manufacturers during the initial introduction of the PLA devices. These fabrication problems have since been overcome. However, the use of the PLA to augment the microprocessor speed has been overshadowed by the emphasis on the inherent speed and performance of the microprocessor alone. While microprocessor speed is important, the hardware control capability achieved by the use of the PLA requires that this device be carefully evaluated for use in the overall microprocessor scheme.

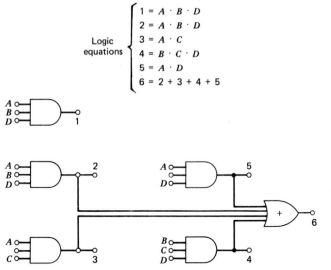

Logic equations
$$\begin{cases} 1 = A \cdot B \cdot D \\ 2 = A \cdot B \cdot D \\ 3 = A \cdot C \\ 4 = B \cdot C \cdot D \\ 5 = A \cdot D \\ 6 = 2 + 3 + 4 + 5 \end{cases}$$

Figure 7.11. Logic diagram for PLA "AND–OR" matrix shown in Figure 7.10.

PROBLEMS

1. When would you use a ROM in your computer?
2. What is the primary use of a PROM?
3. What is the primary use of the PLA?
4. Provide three reasons for using mask-programmable ROMs over field-programmable ROMs.
5. Using the PLA technology, design a circuit to solve the following logic equation:
$$A \cdot (B + C) + C \cdot (B + A)$$
6. Illustrate the resultant field-programmable ROM to hold the following program:

Word 1	1 0 0 1
Word 2	1 1 0 0
Word 3	1 0 1 1
Word 4	1 0 0 1
Word 5	1 1 1 1
Word 6	0 0 0 0

CHAPTER 8

The Central Processing Unit (CPU)

The "brain" behind the microprocessor (or any computer) is the CPU. This unit is normally a composite of several unique computer elements including the arithmetic unit (AU), the logic unit (LU), a stack pointer, a bus unit, a clocking system, and an interface system. These elements are, in the modern microprocessor, located on a single chip (or at most several chips) that interfaces with the RAM and ROM memory units as well as the input and output interface units.

Since the CPU is a composite of the above six computer elements, the individual elements cannot be easily separated from each other without losing their significance within the overall CPU operation. Thus, each element is briefly explained here as to its operation and significance within the CPU architecture and then the operation of the overall CPU examined in detail.

To provide added insight into the design considerations of each of the CPU elements, a generalized discrete circuit (7400 series components) corresponding to the function of the element is described, along with the corresponding unique design features of the discrete circuit whenever possible. However, it must be remembered that while the MOS chip design may differ radically from the discrete circuits described, the functional design will remain similar.

In order to establish a systematic understanding of the CPU and the interaction between the individual elements comprising the CPU, the elements are described in the following order: (a) bus system, (b) two-phase clock and strobe generation system, (c) stack-pointer system, (d) arithmetic logic unit (ALU), and (e) CPU interface system.

THE BUS SYSTEM

The interface between the internal–internal and internal–external elements of the computer and the CPU is maintained by means of a tristate bus system. That is, virtually all data signals from the CPU are transmitted on one set of data lines sharing common input/output terminals with other transmitting/receiver elements of the computer. This system of sharing a common bus, however, could become both confusing and wasteful of power if not controlled by a technique to permit only one transmitted message to be on the line during a single period of time, all other transmitters being disabled to prevent them from loading the transmitted

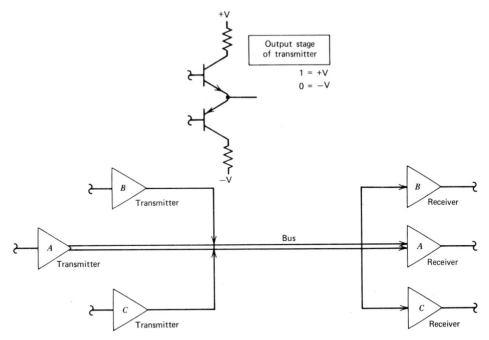

Figure 8.1. Transmission line bus.

signal. This requirement may be met by utilizing a transmission technique called the *tristate transmitter/receiver* circuit.

Figure 8.1 illustrates the parallel nature of a complex common transmission line bus. If the transmitter A is at a level "1" (high output state) while either transmitter B or C are at a level "0" (low output state), a short circuit between the power levels ($+V$ and $-V$) will occur. In addition, the receivers will be unable to read the level of the data being transmitted from transmitter A. Thus, it may quickly be concluded that the parallel operation of two or more transmitters on the same line, without some form of isolation and switching technique, is unfeasible. If, however, a technique of isolation and switching is inserted between the unused transmitters and the common bus, the logic state of the unused transmitters will have no effect on the desired transmitted data. In essence, the only effective transmitter on the line is the one being used, and the system becomes a single transmitter bus line.

The tristate driver (transmitter) (see Figure 8.2) is designed to have three possible output states: (a) voltage high (1), (b) voltage low (0), and (c) high impedance (open). It is during state (c) that isolation from the common data bus occurs. By the use of a timing signal (strobe), a designated transmitter may be connected to or isolated from the common bus transmission line. By using this tristate driver technique, the microprocessor may be fabricated with only one set of common data transmission lines (data bus lines) instead of requiring a unique set of data lines for each individual transmitter/receiver element in the system. The tristate isolation technique accounts for a great savings in interconnection wiring within the microprocessor and permits the fabrication of the single-chip CPU.

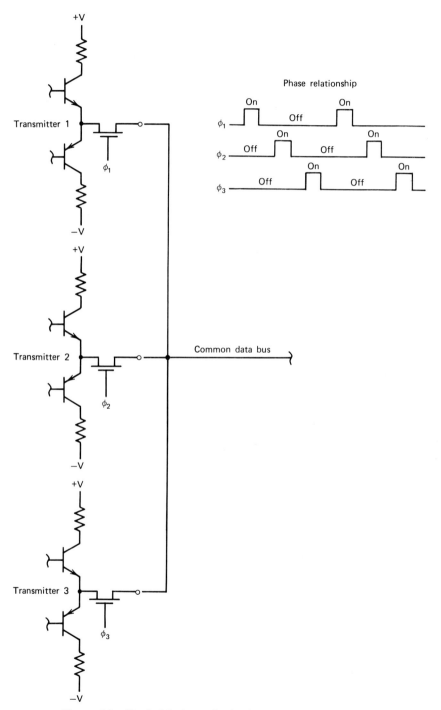

Figure 8.2. Typical logic mechanization of tristate transmitters.

THE MULTIPHASE CLOCK AND STROBE GENERATION SYSTEM

With the introduction of the common data bus interfacing with many tristate driver circuits, the problem of establishing the time interval during which an individual driver will transmit data on the line, versus the time when it will be isolated from the line, becomes important. The solution to this problem gives rise to the basic computer timing sequence.

The computer is operated on a series of timed signals formed by the timing interrelationship of the clock. This clock is generally a symmetrical oscillator that provides a basic (master) timing interval or clock waveform (f_0). This time interval is established to guarantee each sequence of events occurring within the computer will be completed before initiating any subsequent event. Thus, the speed of the basic clock interval is a function of the technology used in the computer-component fabrication as well as the general layout of the computer. The single clock cycle permits two events to be timed, one at clock high-time period and one during clock low-time period.

In general, more than two events must take place to perform a complete operational cycle in the computer. To permit this expanded routine, the number of elements used to generate the timing sequence is increased. The second element introduced into the master timing sequence is a counter (generally a two-count flip-flop counter) which acts to *half* of the fundamental frequency of the symmetrical clock signal. This frequency is referred to as the $f_0/2$ clock. The third element introduced into the timing generation circuitry is a decoder circuit that is utilized to generate a double-phase clock. The phase one clock (ϕ_1) is generated by the logical "AND" signal formed by the (f_0) clock and the $f_0/2$ clock. The phase 2 clock (ϕ_2) is generated by the decoder as the "AND" signal formed by the combination of the f_0 clock and the $\overline{f_0/2}$ clock signal. Figure 8.3 illustrates the timing relationship between these signals, phases 1 (ϕ_1) and 2 (ϕ_2) corresponding to the two phase clocks required to operate the microprocessor. It should be noted that the fundamental clock frequency (f_0) can be generated by either a discrete oscillator circuit, crystal oscillator, or the unique clock generation chips presently available on the market.

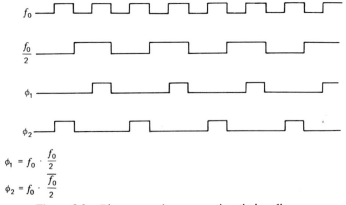

$$\phi_1 = f_0 \cdot \frac{f_0}{2}$$

$$\phi_2 = f_0 \cdot \frac{\overline{f_0}}{2}$$

Figure 8.3.   Phase ϕ_1 and ϕ_2 generation timing diagram.

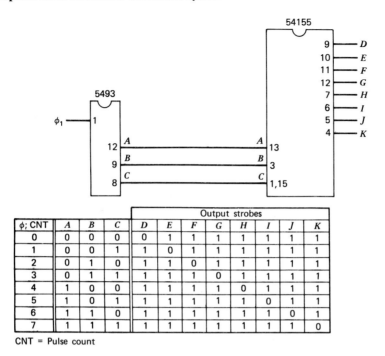

ϕ; CNT	A	B	C	Output strobes							
				D	E	F	G	H	I	J	K
0	0	0	0	0	1	1	1	1	1	1	1
1	0	0	1	1	0	1	1	1	1	1	1
2	0	1	0	1	1	0	1	1	1	1	1
3	0	1	1	1	1	1	0	1	1	1	1
4	1	0	0	1	1	1	1	0	1	1	1
5	1	0	1	1	1	1	1	1	0	1	1
6	1	1	0	1	1	1	1	1	1	0	1
7	1	1	1	1	1	1	1	1	1	1	0

CNT = Pulse count

Figure 8.4.   Strobe generation technique.

Strobes

One of the primary functions of the main timing logic in the microprocessor is the generation of the many strobe (enable) signals required to activate at the proper time the computer's tristate input selectors, memory, and output ports that interface on the common bus system. These strobe signals can be generated in a number of ways. One popular technique is the utilization of a simple 3-bit counter system. Figure 8.4 illustrates this technique. Eight strobe signals are created by combining a 4-bit binary counter (type 5493), an 8-channel demultiplexer (type 54155), and the phase 1 (ϕ_1) timing signal. Figure 8.5 illustrates the timing diagram relating the phase 1 signal (ϕ_1) with each strobe (both primed and unprimed signals) signal.

Each of the signals transmitted on a common data-transmission bus is activated and timed by a unique strobe signal in order to guarantee that two tristate driver circuits will never be turned on simultaneously. In addition, the timing relationships between the read interval and write interval is also maintained by means of the read/write strobe relationship. All timing sequence signals for the computer are formed by the generated strobe signals. Additional strobe signals can be generated from the phase 2 clock utilizing the circuit shown in Figure 8.6. These strobe signals can be further combined by means of discrete digital circuits to form more complex strobe patterns such as shown in Figure 8.7.

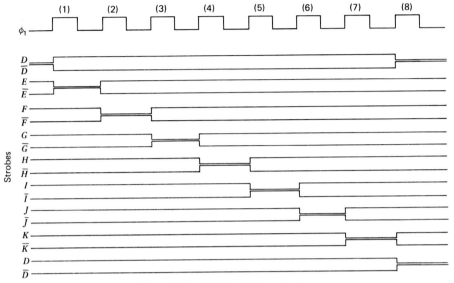

Figure 8.5. Strobe timing relations.

THE STACK-POINTER SYSTEM

The main requirement for the stack pointer is to indicate (remember) the program memory location (normally the location in ROM memory) of the next step in the program sequence. In a small system the stack pointer is the program-control counter and could be implemented by a simple counter that is incremented each time an instruction is completed. Figure 8.8 illustrates a minimum stack pointer/ROM memory-circuit configuration that could be used to control a 256-byte ROM program in a microprocessor.

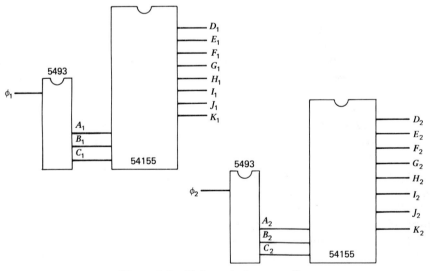

Figure 8.6. Biphase strobe generation.

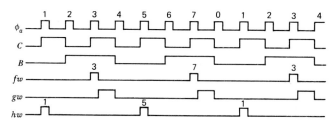

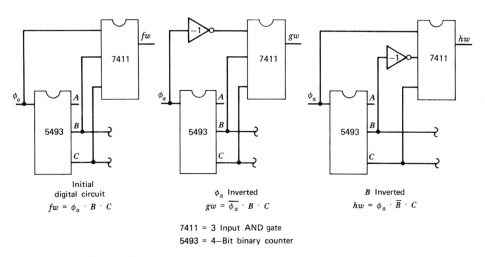

Figure 8.7. Forming complex strobe signals by digital circuits.

In a more complex microprocessor system, the *branch* or *transfer* command is used to change the sequence or location of the program-memory steps. Under these conditions the stack pointer must be provided with a means of saving the original program-sequence location (the last location of the normal program sequence before branching) while loading the stack pointer with the new address location corresponding to the branch or transfer command. The new address location stored in the stack pointer will automatically be incremented at the conclusion of each instruction, until commanded to return to the original program sequence or transfer to another location.

The expanded stack pointer operates in conjunction with a modified parallel-shift register called the *stack,* the register remaining fixed until shifted by the introduction of a branch or transfer command signifying a new program-memory location. When a transfer command is recognized, the shift register is increased by *upward* or *downward* increments, depending on whether the transfer is back to the original program memory location or to a new program memory location. This process is called *nesting* and provides limited branching capabilities for the computer. Transfers move the data in the stack, up or down the stack, one location at a time. This gives rise to the descriptive name LIFO (last in—first out) which is applied to some stack program memories. Figure 8.9 illustrates the movement of the stack and the importants of the stack pointer.

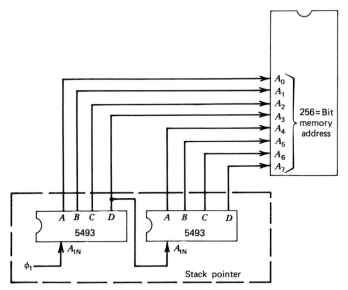

Figure 8.8. Minimum stack pointer to control one page of memory.

THE ARITHMETIC LOGIC UNIT (ALU)

The arithmetic logic unit (ALU) is a combination of the arithmetic unit (the computational element of the microprocessor) and the logic unit (the decisionmaking element of the microprocessor). This combination of mathematical and logical functions gives the ALU a high degree of versatility within a single chip, the relationship between the AU and LU operation of the device being controlled by the state of a single mode select signal.

The forerunner of the modern ALU was the type 54181 bipolar TTL ALU. This bipolar device has the complexity of 75 individual digital gates diffused on a single monolithic chip. The control of this device is maintained by a single-mode control signal coupled with four function-select signals. With this control, the bipolar ALU is capable of performing 16 unique logic functions as well as 16 unique arithmetic functions for a total of 32 different operations. The "truth" table for this bipolar device is shown in Figure 8.10.

The 54181 bipolar ALU interfaces with the external signal data circuitry on three sets of four interface terminals as shown in Figure 8.11. The four input terminals, designated "Word A," can be regarded as the primary data interface to the ALU. The secondary data input terminals to the ALU are designated "Word B." The data output terminals are designated as the "function-output" terminals.

The primary or A input terminals are always tied to the accumulator or A register of the microprocessor CPU. This permits the primary input to remain static (not switched to alternate locations on the data bus) during all operations of the ALU. The bus interconnection between the ALU and accumulator can be switched to permit data to be directed through the accumulator without being arithmetically or logically operated on by the ALU.

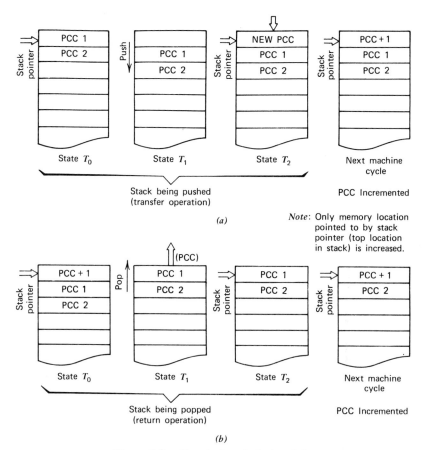

Figure 8.9. The stack and stack pointer.

The secondary input of the ALU is able to interface with the bidirectional bus and, therefore, can directly read any memory location within the computer's RAM/ROM memory system.

The "function output" terminals of the ALU can interface with either the accumulator and/or the bidirectional bus. In the normal microprocessor ALU architecture, all data are placed into the accumulator, the accumulator being connected to the bidirectional bus. The "function-select signal" input is the *operation code* from the program memory and is latched into an external 4-bit register. The accumulator register is a 4-bit latch that holds the operating data during the periods of time devoted to setting up the ALU for the next arithmetic or logic operation.

Figure 8.12 illustrates the discrete ALU circuit diagram for performing the microprocessor ALU operation. During each clock cycle an operational decision is made, based on the program ROM bit sequence, to either: (a) input data from the RAM to the accumulator (which would destroy the existing information/data in the accumulator), (b) input data from the RAM to the secondary input, (c) out-

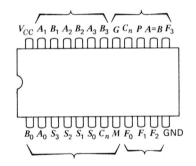

V_{CC} A_1 B_1 A_2 B_2 A_3 B_3 G C_n P $A=B$ F_3

B_0 A_0 S_3 S_2 S_1 S_0 C_n M F_0 F_1 F_2 GND

DM54181/DM74181
PIN DESIGNATIONS

DESIGNATION	PIN NOS.	FUNCTION	DESIGNATION	PIN NOS.	FUNCTION
$\bar{A}_3, \bar{A}_2, \bar{A}_1, \bar{A}_0$	19, 21, 23, 2	Word A Inputs	A = B	14	Comparator Output
$\bar{B}_3, \bar{B}_2, \bar{B}_1, \bar{B}_0$	18, 20, 22 1	Word B Inputs	$\bar{P}$	15	Carry Propagate Output
$\bar{S}_3, \bar{S}_2, \bar{S}_1, \bar{S}_0$	3, 4, 5, 6	Function-Select Inputs	C_{n+4}	16	Carry Output
C_n	7	Carry Input	$\bar{G}$	17	Carry Generate Output
M	8	Mode Control Input	V_{CC}	24	Supply Voltage
$\bar{F}_3, \bar{F}_2, \bar{F}_1, \bar{F}_0$	13, 11, 10, 9	Function-Outputs	GND	12	GROUND

TABLE OF LOGIC FUNCTIONS

FUNCTION SELECT				OUTPUT FUNCTION	
S3	S2	S1	S0	NEGATIVE LOGIC	POSITIVE LOGIC
L	L	L	L	$F = \bar{A}$	$F = \bar{A}$
L	L	L	H	$F = \overline{AB}$	$F = \overline{A+B}$
L	L	H	L	$F = \bar{A}+B$	$F = \bar{A}B$
L	L	H	H	$F = $ Logical 1	$F = $ Logical 0
L	H	L	L	$F = \overline{A+B}$	$F = \overline{AB}$
L	H	L	H	$F = \bar{B}$	$F = \bar{B}$
L	H	H	L	$F = \overline{A \oplus B}$	$F = A \oplus B$
L	H	H	H	$F = A+\bar{B}$	$F = A\bar{B}$
H	L	L	L	$F = \bar{A}B$	$F = \bar{A}+B$
H	L	L	H	$F = A \oplus B$	$F = \overline{A \oplus B}$
H	L	H	L	$F = B$	$F = B$
H	L	H	H	$F = A+B$	$F = AB$
H	H	L	L	$F = $ Logical 0	$F = $ Logical 1
H	H	L	H	$F = A\bar{B}$	$F = A+\bar{B}$
H	H	H	L	$F = AB$	$F = A+B$
H	H	H	H	$F = A$	$F = A$

With mode control (M) HIGH: C_n irrelevant
For positive logic: logical 1 = HIGH Voltage
              logical 0 = LOW Voltage
For negative logic: logical 1 = LOW Voltage
              logical 0 = HIGH Voltage

TABLE OF ARITHMETIC OPERATIONS

FUNCTION SELECT				OUTPUT FUNCTION	
S3	S2	S1	S0	LOW LEVELS ACTIVE	HIGH LEVELS ACTIVE
L	L	L	L	F = A minus 1	F = A
L	L	L	H	F = AB minus 1	F = A+B
L	L	H	L	F = $A\bar{B}$ minus 1	F = A+$\bar{B}$
L	L	H	H	F = minus 1 (2's complement)	F = minus 1 (2's complement)
L	H	L	L	F = A plus $(A+\bar{B})$	F = A plus $A\bar{B}$
L	H	L	H	F = AB plus $(A+\bar{B})$	F = $(A+B)$ plus $A\bar{B}$
L	H	H	L	F = A minus B minus 1	F = A minus B minus 1
L	H	H	H	F = $A+\bar{B}$	F = $A\bar{B}$ minus 1
H	L	L	L	F = A plus $(A+B)$	F = A plus AB
H	L	L	H	F = A plus B	F = A plus B
H	L	H	L	F = $A\bar{B}$ plus $(A+B)$	F = $(A+\bar{B})$ plus AB
H	L	H	H	F = $A+B$	F = AB minus 1
H	H	L	L	F = A plus A†	F = A plus A†
H	H	L	H	F = AB plus A	F = $(A+B)$ plus A
H	H	H	L	F = $A\bar{B}$ plus A	F = $(A+\bar{B})$ plus A
H	H	H	H	F = A	F = A minus 1

With mode control (M) and C_n low
†Each bit is shifted to the next more significant position.

Figure 8.10. DM 54181 truth tables (Courtesy of National Semiconductor).

put data from the accumulator to the RAM, or (d) perform an arithmetic/logic function.

A simple program to add two numbers together, the first a constant stored in ROM memory and the second, previously stored number in RAM, may be performed as follows:

Instruction	ALU Operation
1. Load accumulator from memory.	Step 1. ROM memory is called onto the bidirectional bus by the control program "read" signal during strobe "a" time (T_1, T_2) and the first number is inputted into the accumulator. At T_2 time (strobe b) the accumulator locks this data into its register. No other action is taken.
2. Add memory to accumulator.	Step 2. RAM memory is called onto the bidirectional bus and during strobe "e" time (T_3-T_6), the second number is locked into the secondary input terminals of the ALU. At T_1 time the ROM program bits controlling the ALU function-select operation are inputted into the 4-bit latch, which is subsequently locked into the latch at time T_4 (strobe d). The ALU automatically performs the "add" operation requested by the ROM program. After all transients have died out, strobe c reads the output of the ALU (T_4-T_7) onto the accumulator input lines. At T_6 the accumulator locks these resultant data $(A + B)$ into its register (strobe b).
3. Load memory from accumulator.	Step 3. RAM memory is called onto the bidirectional bus by the control program and is written into during strobe f time (T_7).

A timing diagram for the "ADD" program just described is shown in Figure 8.13. While this timing sequence and "ADD" program is an illustration of what may be performed by the simple ALU circuit, it can easily be seen that the alternatives to the program sequences are boundless and depend only on the complexity

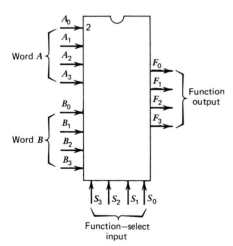

Figure 8.11. ALU interface lines.

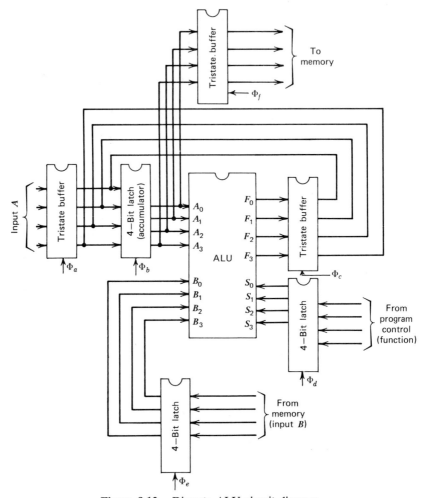

Figure 8.12. Discrete ALU circuit diagram.

and variations introduced in the timing, strobe generation, and bus configurations designed into the overall microprocessor system. It is this added complexity that has been provided in the modern CPU chip to transform it into an operational computer rather than the simple ALU chip just examined.

THE CPU INTERFACE

The interface requirements for the CPU to the external microprocessor elements have already been superficially discussed in reference to ALU structure. The interface of the CPU, under the command of the ROM program, not only forms a buffer between the CPU and all external electronics, but also can perform signal conditioning and multiplexing interface signals to direct the signal flow to and from the ALU and memory elements of the microprocessor. In general, the interface is formed by combinations of tristate logic registers that hold the input/output signals momentarily while the CPU is being set up for other operations.

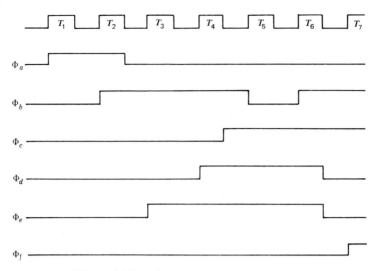

Figure 8.13. Discrete ALU timing diagram.

The main CPU interface is the accumulator that is used to hold all intermediate data. Primary data can neither be inputted to or outputted from the CPU except through the accumulator. The exception to this rule is the direct memory access (DMA) technique, which is employed in several of the newer CPU chips. In general, the accumulator or A register is the central data clearing house for all CPU functions.

The CPU chip provides multiple *on-chip* registers within the CPU chip for use as scratch-pad storage (temporary storage) of data prior to either: (a) writing the data into RAM memory, (b) using the data for subsequent arithmetic or logic manipulation, or (c) outputting the data to an external output register.

Internal data transfer may generally be from any of these internal registers to any other register or to the RAM memory. In most CPU chips, movement of data cannot be made directly to the output registers from the scratchpad registers. To output data which has been stored in a temporary scratchpad register, the data must first be transferred to the accumulator and from there to the output register.

PROBLEMS

1. Describe in detail the tristate bus concept. Describe its major advantages and disadvantages.

2. Design a strobe-generating system to control the transmit time of eight bus transmitters. Use the 8 : 1 (54155) decoder chip.

3. Using the 54181 ALU chip, illustrate the control-function code required to perform the following operations:

      a. $A - 1$ (decrement)   d. $[A + B]$ plus A
      b. $A + B$ e. $A + \overline{B}$
      c. $A \cdot B$ f. $A = B$

CHAPTER 9

The Microprocessor Interface

The input interface-data signals to the microprocessor can be either digital or analog. If the input signal is of digital form, the interface electronics is required to structure or format the input data into an organization that the microprocessor can recognize and operate on. If, however, the input is an analog signal, additional interface electronics are required to transform the analog signal into a digital format that is organized in an acceptable form for the microprocessor to recognize and operate on. (*No digital processor of any form can accept and/or operate on analog input data.*)

The microprocessor can output several formats or forms of digital data ranging from signal flags to data words. To discern the meaning of the output data, additional electronics must usually be provided to translate the output digital data into a meaningful α-numeric listing (typewritten pages, television monitor outputs), or other similar type of restructured output form. This translation may require the reorganization of the microprocessor output data format into one that can be utilized by the output peripheral equipment.

If, however, the output requirement from the microprocessor system is analog in nature, additional interface electronics are required to transform the digital output data into the desired analog signal levels. (*No digital processor of any form can provide output data in an analog form.*)

To understand the basic operation of these digital peripheral interfaces, two forms of digital input equipment are evaluated here: (a) the input keyboard interface and (b) the sequentially sequenced interface. The output digital peripheral interface evaluated is the teletype/television output versus the sequentially accessed receiver.

The analog peripheral interface circuits are evaluated by means of the investigation of the general types of sensors/drivers that may be encountered in both the input and output equipment as well as a general review of the common analog-to-digital (A/D) and digital-to-analog (D/A) conversion techniques utilized in these sensors.

THE DIGITAL PERIPHERAL INTERFACE

The digital peripheral interface has become the most standard form of interface to be used with the digital computer. Two forms of digital peripheral equipment dominate the field, the keyboard input and the sequentially accessed memory source (either tape or card).

The keyboard-input peripheral can range from the single push-button input (toggle switch or momentary push-release switch) to the elaborate teletype keyboard input peripheral capable of typing in any known language or logic symbol.

The sequentially accessed source is found only in the more automated systems. These sources can be as simple as the 80-column IBM card and the punched paper tape, or as complex as the magnetic tape memory system of the open reel, cassette, or floppy disk variety and the dynamic data bus interconnecting the multiple computer systems. While the keyboard peripheral systems are relatively slow, the sequentially accessed input sources can operate at high speeds and may require large quantities of interface electronics to buffer the microprocessor peripheral interface.

The Keyboard

The keyboard is, generally speaking, a human–machine interface. As such, there is a finite rate of speed at which this input can be operated. This operational speed is invariably slower than the minimum permissible input data rate to the microprocessor, which means that the microprocessor can perform other computational activities while pausing to periodically read the input keyboard data.

The simplest form of input keyboard is the single-key (on–off) switch that is operated in response to a signaled condition (i.e., the microprocessor signals an "END OF COMPUTE" state by lighting a lamp on the front panel of the console. When the operator presses the "RESUME COMPUTE" button, the microprocessor reinitializes itself, accepts a new set of variables from RAM memory and performs a new calculation. It should be noted that in this illustration the keyboard controls the function of the machine, not the input data to the machine.) The single-switch keyboard is, however, such a trivial case that it is not perused in the present work.

The second type of input keyboard is the 10-digit, four-function keyboard of the type found in many of the present hand-held calculators on the market today. Several forms of this type of keyboard exist, the original being composed of 15 switches multiplexed together into a binary code. Figure 9.1 illustrates the basic electrical circuit of this keyboard unit.

It can easily be seen that the keyboard can accept at least three additional function keys without increasing the existing electronic logic circuitry. Less expensive keyboards have avoided much of the above electronics by means of specially built printed circuit boards and key fingers. A printed-circuit board for this type of keyboard is shown in Figure 9.2. The basic operation for this keyboard system depends on the slow input rate at which the keys can be depressed, a valid assumption in most cases. If, however, two keys are depressed simultaneously, an error results that has to be cancelled by the operator and the data reentered into the keyboard. In both keyboard forms, the data are normally entered into a parallel-in/parallel-out shift register controlled by the sensing of the leading or falling edge of the input data. The shift register is large enough to hold one or more complete data byte(s).

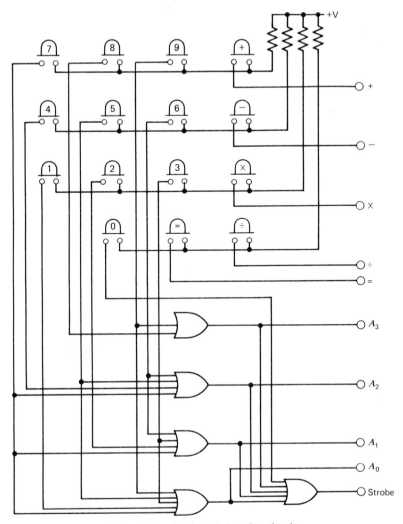

Figure 9.1. Keyboard decoder circuit.

To interface either of these numeric keyboards with the microprocessor, the parallel-in/parallel-out buffer shift register is combined with a second parallel-in/parallel-out temporary storage-shift register within the processor itself. This second shift register (see Figure 9.3) provides for temporary storage of the input data during the time the operator is keying in additional data to be operated on. During this additional period of time, a signal is given to the microprocessor indicating data are available at their input terminals (called an *interrupt* signal) and the microprocessor then pauses from its present operation, inputs the data from the second shift register to memory, and then resumes the operation being performed before the interrupt signal was sensed. By providing an extended parallel shift register for the second register, extended time delays may be handled between: (a) the operator input to the microprocessor peripheral and (b) the processor physically taking the data into the internal memory storage. This option is especially important if the operator desires to check the input data before issuing the data to

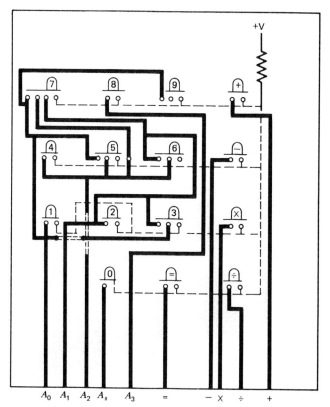

Figure 9.2. Printed-circuit decoded keyboard.

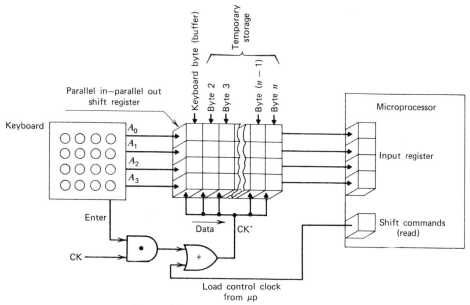

Figure 9.3. Keyboard temporary storage.

the microprocessor. Special devices used to perform the task of the extended storage-shift register are called FIFO (FIRST IN-FIRST OUT) storage registers.

The full teletype peripheral keyboard is becoming increasingly popular as its price decreases. The full keyboard resembles that of a typewriter. Since there is virtually no need for capital letters on the keyboard, each key can perform a double function. Figure 9.4 shows the ASCII code generally accepted for use with these keyboards. However, for special purpose keyboards, any key function may be modified to conform to the internal program requirements of the special purpose microprocessor. Thus, it should be borne in mind that the ASCII code is merely a convention adopted for the teletype communication field and should not be regarded as unchangeable.

Since the full keyboard is composed of 64 symbols, a 6-bit binary code is required to decode the keyboard into an acceptable digital code. In addition, the previously described extended shift register circuit is required to provide temporary storage between the keyboard and the microprocessor. Figure 9.5 shows the ASCII 6-bit code/symbol relationship currently in use for most interface requirements. Figure 9.6 shows the typical electrical interface wiring for the teletype peripheral-to-microprocessor interface using the FIFO shift-register interface.

The second form of temporary storage that may be used to interface the typewriter with the microprocessor is the direct memory address to RAM memory. The major problem with this technique, however, is the continuous addressing that must take place in order to accommodate the continuously changing typewriter input data. To overcome this problem, a moderate amount of external RAM mem-

A	B	C	D	E	F
G	H	I	J	K	L
M	N	O	P	Q	R
S	T	U	V	W	X
Y	Z	1	2	3	4
5	6	7	8	9	0
!	"	#	$	%	&
'	(	)	*	+	,
-	[	]	.	/	:
;	<	=	>	?	@
↑	←	\	SPACE		

Figure 9.4. ASCII Symbols.

ASCII	Symbol	ASCII	Symbol
000000	Space	100000	@
000001	!	100001	A
000010	"	100010	B
000011	#	100011	C
000100	$	100100	D
000101	%	100101	E
000110	&	100110	F
000111	'	100111	G
001000	(	101000	H
001001	)	101001	I
001010	*	101010	J
001011	+	101011	K
001100	,	101100	L
001101	-	101101	M
001110	.	101110	N
001111	/	101111	O
010000	0	110000	P
010001	1	110001	Q
010010	2	110010	R
010011	3	110011	S
010100	4	110100	T
010101	5	110101	U
010110	6	110110	V
010111	7	110111	W
011000	8	111000	X
011001	9	111001	Y
011010	:	111010	Z
011011	;	111011	[
011100	<	111100	\
011101	=	111101	]
011110	>	111110	↑
011111	?	111111	←

Figure 9.5.   ASCII Symbols—6-bit code.

ory may be provided in the peripheral equipment for temporary direct data storage. After a minimum quantity of input data is typed into the peripheral external RAM memory, a transmit code is punched in, locking the keyboard and signaling the microprocessor to input the awaiting information. When the processor has completed the input of all available data, the keyboard is automatically unlocked by an "INPUT COMPLETE" flag signal, permitting additional data to be typed in. Signal lights controlled by the microprocessor could provide visual signals to the typewriter operator. Since the microprocessor is significantly faster than the operator, the time for inputting data is insignificant, appearing to be almost instantaneous to the operator. This operation can be either automatic, semiautomatic, or manual.

The Sequentially Accessed Source

The alternative to the keyboard real-time data input is the use of sequentially accessed data. This type of data input ranges from punched cards to magnetic tape storage and random communication links.

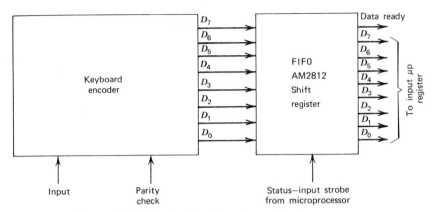

Figure 9.6. Electrical interface—typewriter to processor.

The general description of the sequentially accessed memory source is provided in Chapter 5. The primary advantage of this type of input source is the higher speed machine–machine interface permitted by its use and, as a result, the minimization of the computer interrupt time encountered when inputting data. The input data rates between the data input and processor are now compatible. However, the addition of a data controller is required to maintain the two machines in sync. In general, the operation of the sequential source is forced to be directly controlled by the processor, the source being stopped during each read period while the processor accepts data and then incremented when the processor signals that it read the data into memory.

The actual cycle operates as follows:

1. The sequentially-accessed source advances to a data byte and stops.
2. The data is read photo-electrically and transmitted through a buffer/pulse shaping network to the input register of the processor.
3. The processor accepts the "DATA AVAILABLE" interrupt command and looks for the input data (checks each input register for one containing data). Once the processor locates the input register containing the input data, it reads the data into memory (via direct memory access or via the accumulator–memory minicycle).
4. The processor then issues the "READ COMPLETE" flag signal to the sequentially accessed source controller to advance it to the next data byte, which again initiates the data input cycle.

Since the input data from the sequentially accessed source is directly entered into the input register of the processor, it is obvious that the data must be fully organized in an acceptable format for the CPU to operate on. No reorganization of the data bits is permissible unless the processor has been preprogrammed to perform this reorganization. In larger computers this reorganization is controlled by a unit termed a *compiler*. Most microprocessors are too small to contain a compiler as such, and therefore, must be provided data organized in a form recognizable by the processor.

The digital output peripheral equipment operates in basically the same form as the input peripherals just described, with the exception that everything is reversed.

When the output word is received by the teleprinter or television monitor, it must convert the output byte from a binary code to a code recognizable by the peripheral. The processor places one byte of data into the desired output register and waits for a return signal signifying that the data have been read by the peripheral. Once the data are read and the appropriate signal issued, the processor places the next output byte into the output register and repeats the minor cycle.

During this minor cycle period the peripheral reads the output byte and stores it in a temporary memory (usually a shift register) to await code translation. Once the byte is translated into the correct machine code, it is placed into additional shift registers to be serially outputted (in the case of a telemonitor). Figure 9.7 shows the block diagram of a television-monitor output peripheral.

The sequentially accessed output peripheral also performs its function in reverse to the input sequentially accessed peripheral. The processor controls both the input and output peripheral controllers in order to maintain sync between the two pieces of equipment. The processor outputs a data byte to the sequentially accessed peripheral and waits for the peripheral to read the byte. Once the byte has been read, the processor waits until a signal has been received indicating that the data has been recorded (punched or read into the output tape) by the peripheral equipment. The processor then issues a command to the peripheral controller to increment one space and stop. The processor then, on receipt of a signal that the sequentially accessed output peripheral has, indeed, incremented one space, issues the next data byte to the output register and initiates another output read data cycle.

No translation of the binary output data code is required in the machine–machine interface. Only when the code is read by a human operator is there a need for translating the binary code to one more recognizable by the operator.

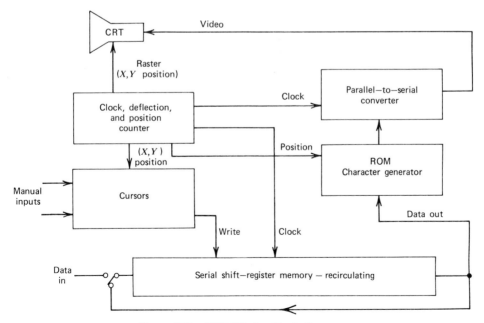

Figure 9.7. CRT Display block diagram.

THE ANALOG INTERFACE

The microprocessor control system utilizes many unique analog sensors as it measures the physical parameters of mother nature. These parameters can range from sensing simple temperature and pressure variations to sensing variations in wind velocity and/or the relative motion between two surfaces. In general, however, the analog measurement reduces to that of sensing temperature, pressure, or rate of flow.

Each of these three forms of measurement is ultimately reduced to an analog magnitude, which, if utilized with any form of digital system, must be transformed into a binary quantity. This is accomplished in an analog-to-digital (A/D) converter. Figure 9.8 shows a typical A/D parallel converter utilizing discrete components.

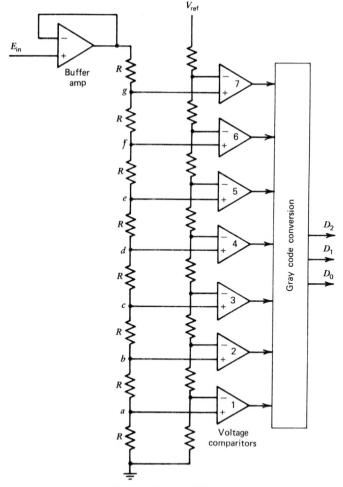

Figure 9.8. A/D Parallel converter.

The Analog-to-digital (A/D) Converter

The input analog signal is normally buffered by a monolithic unity-gain amplifier to permit the converter to maintain a high input impedance. The voltage output of the buffer amplifier is used as the input to a precision attenuation network. The voltage at each level of the attenuation network is compared with the voltage from the reference-level attenuation network through a high-gain voltage comparator. The output of the voltage comparators form a *gray code*. That is, once the desired full-scale input voltage level for the A/D converter has been determined, the reference-bridge voltage is established. When the input voltage is equal to 0.143 V_{fs} (where V_{fs} is the predetermined full-scale voltage for the A/D converter), the output of voltage comparator number 1 will switch from zero (0) to one (1), yielding the output gray code 0000001. When the input voltage level reaches 0.286 V_{fs}, the second voltage comparator will switch, yielding the gray code 0000011. The complete gray code versus input voltage range for the A/D converter illustrated is shown in Table 9.1.

The output gray code from the A/D converter has very little utility unless it is converted to a binary code. This conversion is most easily performed in a gray-code-to-binary converter. This monolithic chip performs the task of transforming the 8-bit gray code into a 3-bit binary code. It can be seen that there are seven voltage comparators used to translate the input analog signal into a 3-bit binary code having a 14.3% granularity between sense levels. To construct an A/D converter of this type to transform the input voltage into an 8-bit output code would require 255 voltage comparators with the appropriate attenuation circuits. A device of this type is very expensive.

A more popular A/D converter, and one significantly less expensive than the converter previously cited, is the successive-approximation A/D converter. This converter uses the technique of comparing the unknown input signal with a precisely generated internal voltage generated by a D/A converter to determine its binary output value. The operation of this converter is as follows:

1. The most significant bit of the internal D/A converter is turned on and the resulting analog voltage compared with the input voltage level. If the feedback

TABLE 9.1. Input A/D Voltage Range versus Output Gray Code

Input Range Normalized to Full Scale	Gray Code	Binary Code
0 →0.143	00000000	000
0.143→0.286	00000001	001
0.286→0.428	00000011	010
0.428→0.571	00000111	011
0.571→0.714	00001111	100
0.741→0.857	00011111	101
0.857→1.00	00111111	110
1.00→>1.00	01111111	111

voltage is less than the input voltage, the switch is left on. If the feedback voltage is greater than the input voltage, the switch is turned off.

2. This procedure is repeated until the D/A converter switches have been sampled from the MSB to the LSB.

When a balance is reached, the D/A converter voltage equals the voltage of the unknown input level and the D/A digital switch value (binary number) represents the output binary value of the A/D converter. This form of A/D conversion is significantly slower than the parallel converter technique but is very attractive in terms of price. A block diagram of the successive approximation A/D converter is shown in Figure 9.9. Converters of 8–12-bit accuracy are common using this technique.

Ideally, the A/D conversion process should be performed within the sensing component by the use of a hybrid circuit technique rather than external to the sensor. By using the hybrid circuit configuration, the manufacturer is able to control the overall interface errors between the transducer and the A/D converter since he can maintain full control over both units. With the introduction of practical hybrid technology, the combining of the sensor–converter electronics within a single chip has become a practical reality. Figure 9.10 illustrates the schematic for a pressure transducer in hybrid form. This transducer has not yet been marketed in a combined hybrid package with an A/D converter.

The Digital-to-analog (D/A) Converter (DAC)

The D/A conversion associated with the output interface of the computer is normally the full responsibility of the microprocessor designer. Figure 9.11 illustrates the discrete D/A converter circuit. The operation of the D/A converter is as described in the following paragraph.

The input digital signal from the computer activates the switches of the DAC from the LSB to the MSB. This permits current to flow through the shunt resistors connected to the -10 V level by means of the digital input being a "1." Each successive stage from the MSB to the LSB will sink ½ of the current magnitude of

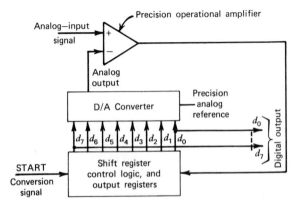

Figure 9.9.   8-Bit successive approximation: A/D converter.

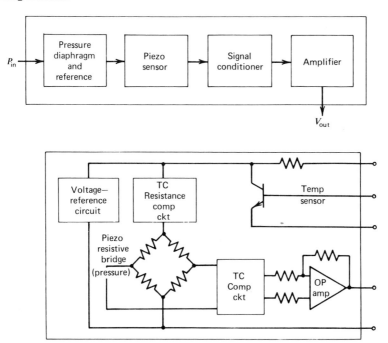

Figure 9.10. Typical pressure transducer.

the preceding stage. The total current drawn from the operational amplifier through the resistor network determines the magnitude of the reconstructed output analog voltage. The accuracy of this output voltage is a direct function of the accuracy of the latter network.

It must be noted that the D/A converter (commonly referred to as the DAC) is extremely difficult to fabricate due to the parameter nonlinearity and tolerances of the discrete components used to construct the circuit. For this reason the DAC

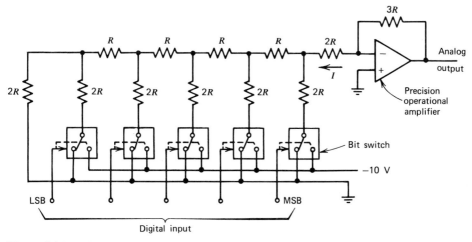

Figure 9.11. Discrete DAC. *Note:* For digital input bit value of 0, bit switch goes to ground. For bit value of 1, bit switch goes to −10V.

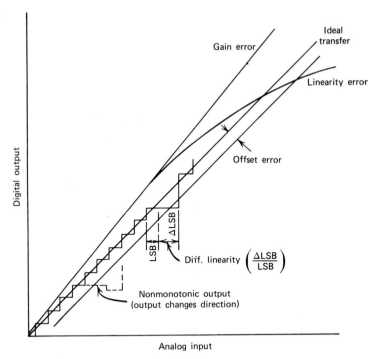

Figure 9.12. DAC errors.

is normally purchased as a hybrid circuit chip. Nevertheless, the D/A conversion may account for the largest errors in the control system while being among the costlier of its components. As a result, it will pay the microprocessor designer to spend extensive time in the evaluation of the DAC circuits on the market before selecting one for his control circuit. Figure 9.12 illustrates some of the primary errors associated with the DAC conversion chip.

The Phase-locked-loop (PLL) Converter

In contrast to the traditional D/A conversion techniques, many analog control elements may be more accurately controlled by the utilization of a phase-locked-loop (PLL) converter. The basic PLL circuit consists of a closed loop system that is servolocked to the input signal frequency appearing at its positive input port (see Figure 9.13). This frequency, when compared to the loop feedback frequency by means of a phase detector, produces an error voltage directly proportional to the difference in phase (frequency) between the input and the feedback frequency. The "error" pulses are then rectified and filtered to produce an analog control voltage (V) which is used to control the voltage-controlled oscillator (VCO) to produce the desired feedback frequency. The loop is said to be in "lock" when the input and feedback frequencies are exactly identical.

When the PLL is used as a digital to analog converter, the demodulated analog control signal becomes the analog output voltage of the converter. Since this ana-

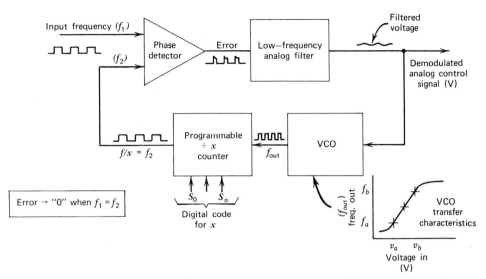

Figure 9.13. Typical phase-lock loop (PLL) configuration.

log control signal is analog, it can be used to directly control many types of analog motors and/or actuators.

The PLL demodulated analog control signal can be directly controlled by two methods: (a) the direct-input frequency simulation technique and (b) the programmable counter technique. The first technique makes use of the normal functional flow of the PLL. If the input signal frequency to the PLL is modified, the "error" voltage is increased (or decreased), resulting in a corresponding change in the demodulated analog control signal. This change in the control signal forces a change in the output frequency of the VCO to maintain the PLL in "lock." The resulting change in the demodulated analog control signal is the parameter of interest.

The second technique used to directly control the PLL demodulated control signal utilizes an externally programmable counter in the feedback loop of the PLL. The VCO raw output frequency is initially scaled to exactly compensate for the frequency division performed by the programmable counter. When, however, the value x of the division counter is modified to $x + \delta$ (by changing the value of the programmable count), an "error" voltage is detected (assuming a constant input frequency) that results in a change in the demodulated control signal. This change results in a shift in the raw VCO output frequency of $x/x + \delta$. The exact change in analog control voltage for either technique is a direct function of the VCO scale factor ($\Delta f / \Delta V$). Thus, effective digital to analog conversion can be achieved using the PLL circuit technique by means of either the direct input frequency simulation or direct VCO frequency division or a combination of the two.

To accomplish the above conversion process by means of the direct input frequency simulation technique requires the use of additional external hardware circuitry. In this technique the microprocessor output control signal (digital byte) is externally formatted into a quasifrequency train, the frequency of the train being established by the $1 - 0$ output pattern generated. This output byte is fed into a parallel in–serial out shift register that is operated in either an open-loop configuration (the data are lost after being read) or a recirculating closed-loop configuration (the output data byte is recirculated until it is updated by a new output data byte) depending on the speed/stability requirements of the servo control system. The resultant magnitude of the demodulated control signal will be in direct proportion to the output byte frequency pattern generated by the microprocessor.

Accomplishing the D/A conversion process by means of the direct VCO frequency division requires the direct coupling of the programming terminals of the feedback division counter to the output terminals of the microprocessor (assuming a dedicated output channel from the microprocessor). The output data byte from the microprocessor establishes the value of the programmed division $x + \delta$ and the resultant change in the demodulated control signal.

The block diagram of a microprocessor/PLL motor speed control is illustrated in Figure 9.14. In this illustration the motor controls the rate of flow (pumping rate) of the fluid from the storage tanks to the utilization area A. The rate of flow of the fluid is to be controlled by the following external parameters: (a) time, (b) temperature, (c) demand, (d) remaining supply of fluid, and (e) rate of resupply of fluid.

Once the microprocessor has sampled each of the five input parameters and calculated the desired motor speed based on these parameters, the pump motor speed must be physically modified. Two microprocessor output data channels (ports) are utilized to perform this task in order to maintain fine control over the motor-speed variations. Output port number 1 (the primary control) utilizes the programmable counter conversion technique to modify the VCO feedback-division ratio within the PLL feedback loop while output port number 2 (the secondary control) is

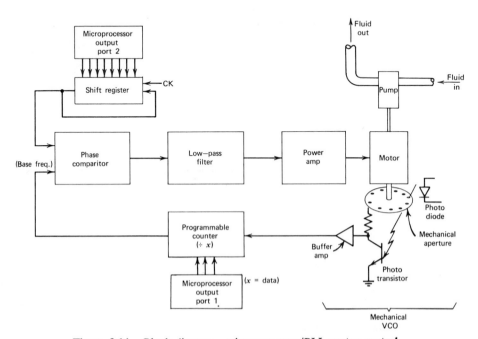

Figure 9.14. Block diagram—microprocessor/PLL motor control.

used for direct input frequency simulation as required. The resultant variation in the demodulator control signal voltage is applied to the direct-current motor winding to directly control the pump speed.

It can be seen that the motor speed is utilized to control the mechanical VCO formed by the mechanical aperture and photo diode/transistor combination. Since the speed (r.p.s.) of the motor is directly proportional to the input voltage to the motor and the output frequency (p.p.r.) is directly proportional to motor speed, the resulting scale factor of the mechanical VCO is $\Delta f / \Delta V$, which is identical to its electronic counterpart. In other words,

$$\left[\frac{\Delta \text{r.p.m. (r.p.s.)}}{\Delta V \text{ (Volts)}} \times \frac{\Delta f \text{ (c.p.s. or Hz)}}{\Delta \text{r.p.m. (r.p.s.)}} = \frac{\Delta f \text{ (c.p.s. or Hz)}}{\Delta V \text{ (Volts)}} \right]$$

The remaining operation of the PLL converter is as previously described.

As indicated by this illustration, the overall peripheral interface is limited only by the imagination of the designer. This section only touched the tip of the iceberg with respect to the peripheral potentials. It is, however, the sphere of activity where the analog engineer will have great opportunity to demonstrate both his inventive capability and his design expertise. As the microprocessor begins to find its way into the dedicated control system market, the A/D and D/A conversion technology will need to expand far more rapidly than the microprocessor field which employs it in order to meet the demands of the marketplace.

PROBLEMS

1. Design the interface (using block diagrams) for interfacing a microprocessor with a solar light sensor. The output of the controller will turn on the lights in the building at dusk and will turn them off at predetermined periods of time. The processor will also control the heat to each room, lowering the heat after dusk and increasing it prior to dawn.

2. Design the interface unit for tracking the sun as it moves during the day and its variations during the seasons. This interface may be used to control a solar platform.

3. Design an interface for direct communication between yourself and an electro-mechanical robot via a microprocessor. The robot can sense temperature and light as well as perform three-dimensional movements.

4. Design the interface (block diagram) required to develop an environmental undersea capsule for marine study. Assume the capsule has only a finite air supply and must be operated in the same fashion as a diver's aqualung. Determine the input parameters to the processor as well as the form of output parameters to the divers.

5. Design the required interface for a complete security system for safeguarding a bank (or other secure agency) against intrusion. Some individuals will have 24-hr. ingress and egress, while others will be permitted entrance only during certain hours. The general public will have no access to the secure area.

CHAPTER 10

Programming The Microprocessor

Programming the microprocessor is the art of directing the processor hardware to perform the logic and/or arithmetic steps necessary to solve a mathematical equation or provide linear control equilibrium by means of internally stored instruction commands. The writing and sequencing of these instruction commands is called *programming*. Since programming change or alteration does not entail physically changing the hardware associated with the microprocessor or its external sensors, this programming function is referred to as *software*. It should be noted that a change in the microprocessor program will require a modification to the ROM program existing in a microprocessor that *is* a device hardware change and results in a minor impact on the processor hardware. The term *software,* however, is a carry-over from the larger computer installations in which the ROM program was stored in a nonvolatile RAM memory rather than a rigid ROM memory. For this reason, PROM memories are normally used in place of ROM memory during the software phases of program development in order to minimize hardware changes when program changes are required.

THE PROGRAMMING LANGUAGES

Programming can be performed at one of three unique language levels: (a) the machine language level, (b) the assembler language level, and (c) the higher language level. However, the three programming levels are not to be regarded as alternates to each other, for they serve distinctly different needs within the spectrum of programming.

Machine-level Language

The machine-level language is the most basic of all programming languages. With this language the programmer "talks" directly to the processor (machine) as shown in Figure 10.1a. This programming level is often regarded as the most difficult to perform since it requires the programmer to think like the machine functions, that is, in extremely simple terms. For this reason many programmers often shun this level of machine interface, arguing that machine programming is too time consuming.

106

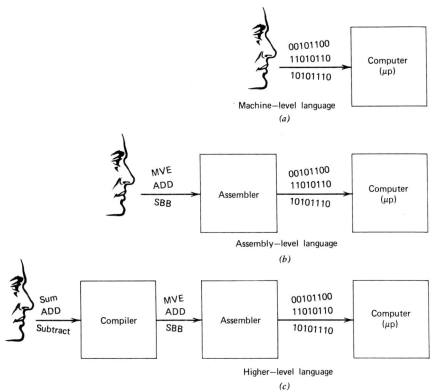

Figure 10.1. Programmer–computer language interface.

The fact is that machine-level programming, while using notations unfamiliar to the beginning programmer, provides the programmer complete control of each sequence of steps that the machine performs. This control in turn provides the programmer with the capability of program manipulation, which will ultimately result in a maximum usage of machine memory with the resultant minimum hardware cost and maximum profit margin in the marketplace.

Assembler-level Language

The second level of programming language separates the programmer from the machine by the use of an assembler, a machine that translates assembler language into machine language (Figure 10.1b). The great advantage of assembler language over machine language is that the assembler language looks more like English, resembling every-day speech. As a result, the use of assembler language places less of a burden on the programmer to learn and understand a "new" language. This "advantage" is not without cost, however. The initial cost is a dollar expenditure for the software program as well as for resident memory in the microprocessor to hold the assembler language. The second cost is a decreased efficiency in the usage of memory due to the lack of control that the programmer has on the actual machine program. This inefficient use of memory results in recurring costs of un-used or redundant memory locations each time a system is built. The summation of these two costs renders the utilization of assembler level programming question-able for the dedicated processor application.

Higher-level Language

The final level of programming language is that of the higher level language (Fig-ure 10.1*c*). This language level is designated by many titles such as FORTRAN, COLBOL, BASIC, and PL/1, as well as several dozen more names. Each of these higher level languages are unique, being composed of different symbols and func-tions. The similarity between these is that each was developed to fulfill the need for a nonexpert programmer to program in a language identical to his professional jargon. Thus, FORTRAN and BASIC are filled with the jargon of the scientific community, employing scientific notations and subroutines in an equation like programming format. In contrast, COLBOL is uniquely formulated to meet the needs of the business community, emphasizing interest rates, percentages, and the business jargon.

The primary cost of this uniqueness in programming language is that the human programmer is forced further away from direct control of the machine, now speak-ing to a mechanical compiler/assembler that talks to the computer in machine language. The positive side, however, is that the nonskilled programmer can write highly sophisticated programs without laborious training in computer programming, since the compiler/assembler can be modified to interface with many machines without significantly changing the programming requirements from the program-mer's viewpoint. The greatest disadvantage of higher language programming is that the memory utilization of these programs is often less than 50% of the available memory space, leaving vast areas of memory locations blank. (This poor memory efficiency is due to the program generalizations necessary to interface between programmer and machine.)

For large computer systems such as the IBM general-purpose computers com-mon in many businesses, such inefficient use of costly memory is justifiable for three reasons. First, the general-purpose computer is intentionally built with a massive memory realizing that the higher-level programming is inefficient in mem-ory usage. Secondly, the cost of training each individual scientist or businessman using the machine to be a proficient machine-level language programmer would be wasteful of valuable professional time as well as a gross misuse of their creative energy, a cost far in excess of that incurred by inefficient use of computer memory. Finally, the general-purpose computer is programmed *continually* as contrasted with one-time programming for the dedicated controller, justifying the use and inefficiency of the higher-level language. Except for the programming of the general-purpose machine, however, the use of higher-level language programming is very suspect in terms of its overall economic costs.

Of the three programming levels, the dedicated microprocessor controller pro-gram should be written in the lowest-level programming language possible. The machine-level language is preferred over the assembly-level language due to the absolute control that the programmer exerts over each element of the program, a control that is lost when the supplemental assembler usurps the decisionmaking control of the programmer. The use of higher language programming should be reserved for the general-purpose machine for which it was initially developed.

The machine-level programming of the microprocessor (as with any other computer not utilizing a higher-level language) is totally dependent on the unique machine being programmed. Learning to program one microprocessor in machine language will provide insight into general microprocessor programming (all machines are similar in architecture) but will not be ultimately useful without the further study of the unique commands of the machine used. The generalized microprocessor is yet to be developed and, as a direct result, there is no generalized microprocessor programming language.

The Intel 8008 microprocessor was the first generally accepted 8-bit microprocessor on the market and is most readily available to the engineer for evaluation. This makes the Intel 8008 microprocessor an ideal system to learn to program. It must be remembered that the Intel 8008 microprocessor has several basic deficiences in its capability since it was the first 8-bit processor on the market. As the forerunner in a new field, it is more complicated to program than the newer 8-bit microprocessors appearing on the market today, including the new Intel 8080 second-generation processor and the Motorola 6800 processor. A summary of the Intel 8080 and Motorola M6800 microprocessor system programming commands is provided in Chapter 15.

THE FLOWCHART PROGRAMMING TECHNIQUE

The initial programming effort, regardless of the programming language used, is a function of the problem being solved. The solution (called a *program* or *algorithm*) is composed of the total sequence of steps required to transform the available data into the desired output solution. However, the scope of most solutions, even a simple one, is far too complex to enable simultaneous visualization and comprehension. In addition, the genius of the microprocessor rests in its ability to make logical choices and to branch to other portions of the program routine based on those choices. With this level of complexity confronting the programmer, a technique of creating a roadmap of the algorithm must be established lest one part of the program be forgotten while concentrating on another.

The technique of creating the roadmap for the program is called *flowcharting*. It is a graphical display of the individual steps being solved in the solution of the algorithm. Two levels of flowchart can be developed: (a) the logic flowchart and (b) the machine-function flowchart. The logic flowchart is constructed on the programmer's logical thought pattern, each block constituting a logical thought. A block in this diagram may be a single thought such as "add $A + B$" or a complex thought such as "solve $\sqrt{x^2 + y^2}$."

The machine-function flowchart differs from the logic flowchart in that each block can contain only one machine operation or step in order to guarantee that each of the instruction codes will be included in the final program. D. D. McCrocken, in his book, *Digital Computer Programming* (Wiley, New York,

1957), comments that such a detailed flowchart can be dispensed with when programming in higher language codes since the detail machine functions are most often covered by the assembler and not the programmer. This is not the case when programming the computer in machine language. At the machine level a complete machine-function flowchart will be required for each logical step the programmer makes. Thus, once the complete flowchart has been completed, a full picture of the programmer's thought process in reaching a solution to the problem may be followed. This capability is especially important in trouble shooting (called *debugging*) the program if it fails to produce the expected results. In addition, while it can be very difficult to follow the sequence of thought represented by a random group of numbers (operation codes) signifying various machine instructions, the flowchart provides a graphic step-by-step explanation of the decisions being made by the computer. It is demonstrated later that the translation of these step-by-step machine operations into machine-language instructions is extremely simple once the flowchart is completed. Debugging and verification of the resultant computer program by means of the flowchart is imperative.

Since there must be communication between individual programmers as well as between the programmer and machine, the essentials of flowchart symbology and notations should be standardized. Again, we generalize the standard of flowchart symbology set forth by McCrocken as the one used in the present book.

Flowchart Notations

The system of flowcharting necessary for use in the creation of a microprocessor algorithm will require, as a minimum, symbols to denote the following:

1. The various functions to be performed.
2. Changes in the sequence of calculations being performed as a result of logical or mathematical decisions.
3. Changes in the sequence of calculations being performed as a result of modifying or creating instructions via the internal program.

Based on the above minimum information requirements, the function box, choice box, connector, assertion box, and summation block are defined as follows.

The Function Box

The box of Figure 10.2 contains the description of the function (detailed machine function) that is to be performed at that point in the program. The function box

Figure 10.2. The function box.

has only one entry and one exit. For example, the function (add: $A + B$) is a single machine function that does not involve any decision or choice of exit on the part of the machine. The next step in the program is found in the next function box.

The Choice Box

The choice box (Figure 10.3) indicates the two possible paths a program may take based on the machine operation performed. As can be seen, the choice has a single entry and a double exit.

   The choice box is normally used for either a "jump" or "call" operation and should designate which condition causes the program to exit from exits 1 or 2. A machine operation may call for the program to jump to path 1 if the condition flip-flop "Z" is in the high state (perhaps indicating an equality between two numbers) or jump to path 2 if the condition flip-flop "Z" is in the low state (indicating inequality). The flowchart representation of the choice box is as shown in Figure 10.4.

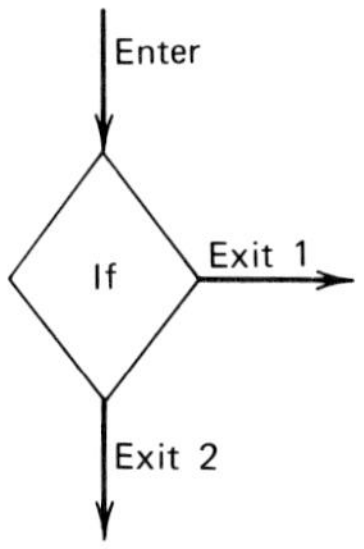

Figure 10.3.   The choice box.

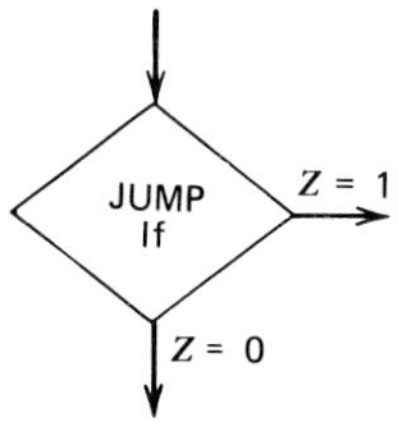

Figure 10.4.   Choice box (JUMP if $Z = 1$).

The Connector

A connector is represented as shown in Figure 10.5. Its primary use is in indicating how to join the signal flow(s) in complex flowcharts where it is not possible to create an unbroken chain. A flowchart signal flow may cease (break) at the left-hand circle and be continued at a different location at the right-hand circle. The use of the connector is especially valid when creating a single subroutine that

 **Figure 10.5.** The connector.

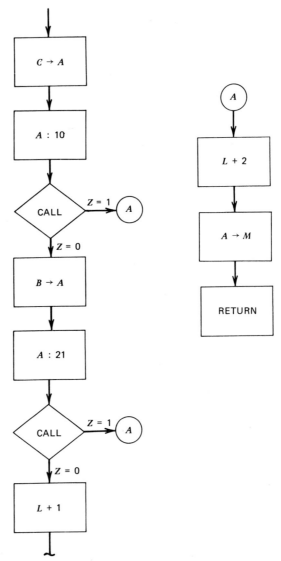

Figure 10.6. Use of connector for multiple use of subroutine.

may be activated from several locations in the program. Such a flowchart representation could appear as shown in Figure 10.6.

The Assertion Box

The assertion box is shown in Figure 10.7. This is a message box for the programmer to make notes in and has no effect on the program. If the assertion box is used in sufficient numbers and at the appropriate locations, it can enhance the value of the flowchart more than any other single device. This assertion box will transform a "cold" flowchart (one that the programmer has not seen for several months or perhaps has never seen) into a rational flowchart that is easily followed. For example, suppose the flowchart called for the data contained in input register 1 to be moved to register B (a computer input command). The flowchart function box would merely contain the input command (input $1 \rightarrow A$) and register to register move command ($A \rightarrow B$) without reference to what data is contained in input register 1. To refresh the programmer's memory that at this point the input register 1 contained the patient's heartrate, an assertion box would be added indicating this information. Figure 10.8 illustrates this usage.

The Summation Block

The summation block defines a junction of two or more signals that enter into one of the flowchart boxes. There is no logic function described by the summation block since it is used only as a graphical aid in the flowchart processes. It is symbolically shown in Figure 10.9. It should be noted that the summation block could be replaced by the connector symbol with very little impact in the overall

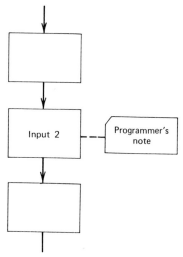

Figure 10.7. The assertion box.

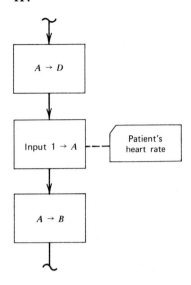

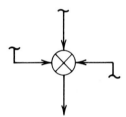

Figure 10.8.   Illustration of use of assertion box.

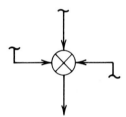

Figure 10.9.   The summation block.

clarity of the flowcharting process. However, for a chain flowchart as seen in Figure 10.10, the use of the summation block is extremely effective in demonstrating the use of the logical loop.

An adequate number of symbols have now been defined for use in developing a flowcharting system for creating machine-level programs. The detailed operational codes to be written in each box is provided in Chapter 11. A final generalized rule to follow in flowcharting is that the programmer can write anything he desires within the flowchart provided that it is meaningful. Some abbreviations, however, should be used to denote the more commonly used machine functions in order to help speed the flowcharting activities. These abbreviations and their meanings are illustrated in Table 10.1.

Machine-level Flowcharting

The translation of the programmer's thoughts into a logic flow diagram is merely the beginning of machine-level programming. Once a valid logic flow diagram has been developed, the logic flow must be translated to a machine operation flow

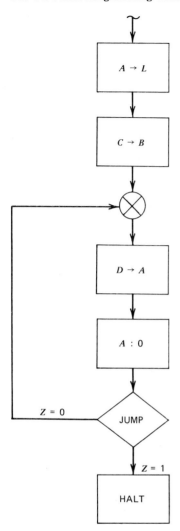

Figure 10.10. Illustration of use of summation block in a program loop.

diagram corresponding to each operation that the microprocessor must perform in order to complete each logical step. If, for example, the logic flowchart box requires the solution to the equation, $C + D + E = A$, the corresponding machine operation flow diagram will show the machine solution to each step of the desired equation. Thus, the first operation is to move C from its memory location to the accumulator. The second operation will ADD the value D to the value in the accumulator yielding $(C + D)$. The final step is to add E to the value in the accumulator yielding the desired result, $(C + D + E)$. The machine-operation flow diagram corresponding to this solution is shown in Figure 10.11. The machine-operation flowchart is constructed using the same functional boxes as utilized in the construction of the logic flow diagram. In general, however, there will be one or more machine-operation flowchart boxes for each logic flowchart box.

TABLE 10.1. Flowchart Abbreviations

Data Movement

1. $A \rightarrow B$ A replaces B
2. $A \leftarrow B$ B replaces A
3. $\overrightarrow{A}$ A rotated to right one bit
4. $\overrightarrow{A} c$ A rotated to right one bit through the carry flip-flop
5. $\overleftarrow{A}$ A rotated left one bit
6. $c \overleftarrow{A}$ A rotated left one bit through the carry flip-flop

Increment/Decrement Functions

7. $A + 1 \rightarrow A$ A increased by 1
8. $A - 1 \rightarrow A$ A decreased by 1

Arithmetic/Logic Functions

9. $A + B \rightarrow A$ A added to B replaces A
10. $A - B \rightarrow A$ A minus B replaces A
11. $A \cdot B$ or $A \wedge B$ AB product
12. $A \veebar B$ Exclusive—OR A,B
13. $A \vee B$ Inclusive—OR A,B
14. $A:B$ Compare A to B $(A - B)$

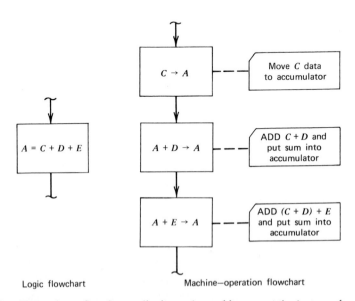

Logic flowchart   Machine—operation flowchart

Figure 10.11. Comparison flowcharts (logic and machine operation) to solve equation $A = C + D + E$.

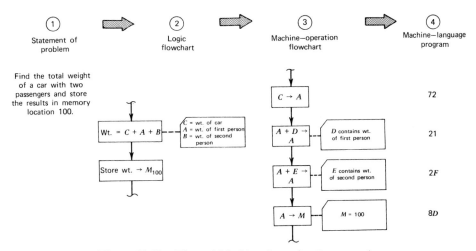

Figure 10.12. Flow of Machine Operation Programming.

Machine-level Programming

The final step in machine-level programming is the determination of the machine-operation code for each machine operation shown in the machine-operation flow-chart. The machine-language program is written in the same sequence as the machine-operation flowchart, one operation code for each flowchart box. The entire programming sequence is shown in Figure 10.12.

Figures 10.13–10.15 provide three examples of programming to aid in understanding the relationship between the logic flowchart and the machine-operation flowchart. In each example the solution to the problem is formatted in four distinct sections. The first section is a block diagram of the microprocessor interface and internal register utilization. (This block diagram is part of the programmer's initial thought process and defines the movement of data into and out of the machine as well as any internal registers having a dedicated usage.) The second section details the logical thought process used to define the program/solution. Section three details the logic flowchart while section four details the corresponding machine-operation flowchart.

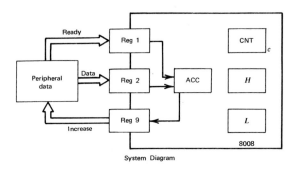

System Diagram

Thought Process	Logic Flowchart	Machine-operation Flowchart	Program

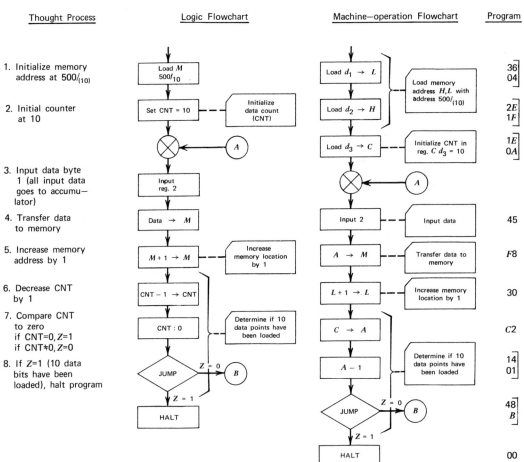

Figure 10.13. Example 1. Input 10 data bytes into register 2 and place these into RAM memory beginning at memory location $500/_{(10)}$.

118

9. Increase tape sequence to next data byte (output d_4 to register 9)

10. Data ready signal (FF at input 1) denotes new data byte is locked into register

11. Computer acknowledges data are ready to be read by returning to main program A

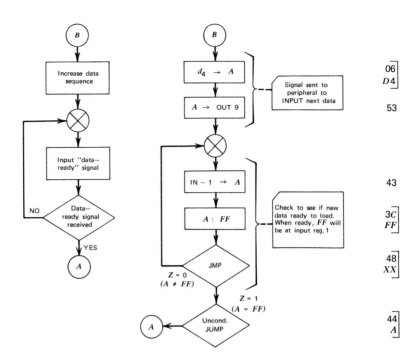

Figure 10.13. (continued)

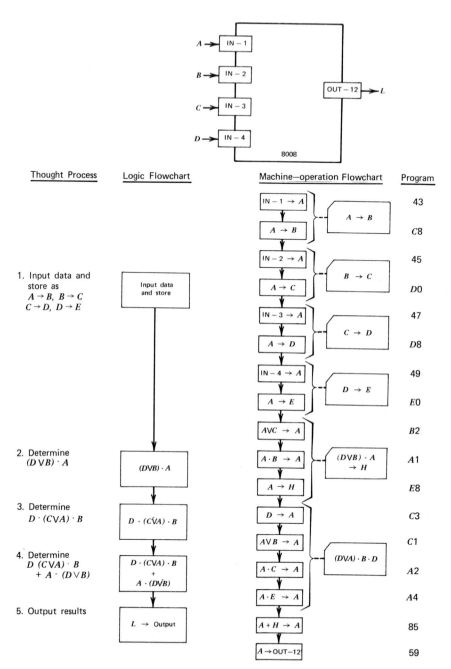

Figure 10.14. Example 2. Solve the logic equation $[A \cdot (B \lor D) + [D \cdot (C \lor A) \cdot B] = L$.

120

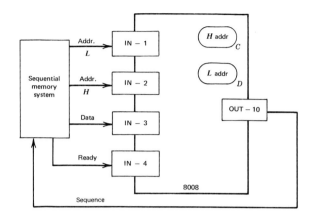

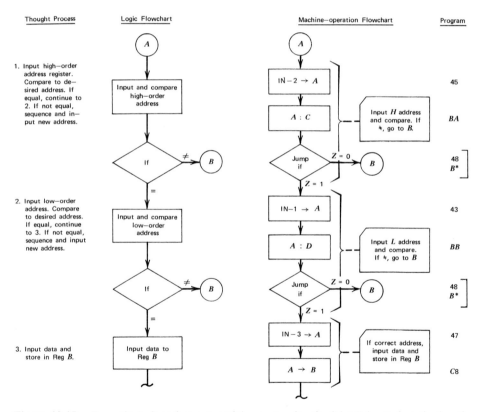

Figure 10.15. Example 3. Search a sequential memory for the "data" located at the location designated in registers C (H addr.) and D (L addr.). Input the "data" to register 3. The C and D register values have been established by a previous subroutine.

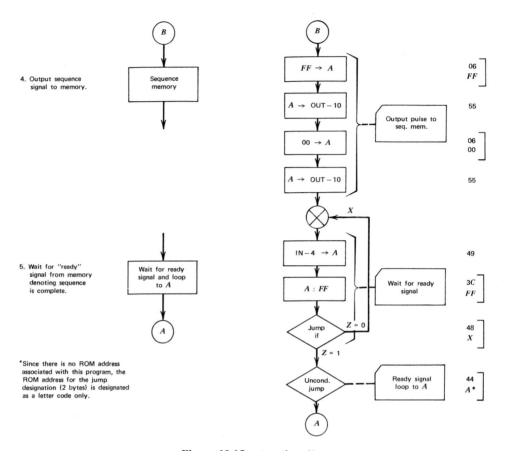

4. Output sequence signal to memory.

5. Wait for "ready" signal from memory denoting sequence is complete.

*Since there is no ROM address associated with this program, the ROM address for the jump designation (2 bytes) is designated as a letter code only.

Figure 10.15. (continued)

PROBLEMS

1. Construct the logic flowchart to solve the equation

$$x = 3y + 7z - k$$

   Transform the logic flowchart into a generalized machine-level operation flow-chart to perform this task.

2. A block of data is to be moved from one block of memory locations (M_x to M_{x+d}) to a new block of memory locations (M_y to M_{y+d}). Develop the logic flowchart and generalized machine-level operation flowchart to perform this task.

3. Develop the logic flowchart to multiply two numbers together. Break the flow into the smallest logic increments possible.

4. A cruise range indicator is required for automotive use. The input sensors will provide data on: (a) remaining gas supply in gallons (input register 1), (b) present speed in miles per hour (input register 2), and (c) present fuel consumption in gallons per hour (input register 3). Develop the logic flowchart required to determine the cruise range of the automobile.

5. A matrix of $X \times Y$ data locations (television screen raster) is to be swept in conjunction with a corresponding data byte (brightness). The X location is outputted at output register 15 while the Y location is outputted at output register 16. The corresponding data-byte magnitude is outputted at output register 17. Assume that the entire matrix data-byte magnitudes have previously been stored in a block of memory beginning at location 256_{10}. Develop the logic flowchart and generalized machine-operation flowchart to perform this task.

CHAPTER 11

The Intel 8008 Machine Language

The programming of a microprocessor is uniquely tied to the organization of the processor ALU/control chip. Figure 11.1 provides an organizational layout of the Intel 8008 microprocessor chip with particular emphasis to the "on-chip" registers utilized by the processor. Figure 11.2 shows the basic instruction cycle of the 8008 and the basic timing relationship creating these cycles. The basic instruction cycle consists of five time periods, which are generated by an external chip TTL logic decoder, which is controlled by the S_0, S_1, S_2 state lines. The 8008 utilizes more operation periods than the generalized computer system examined in Chapter 2 because it utilizes two clock periods to perform certain steps such as addressing to compensate for its short 8-bit word length and its pin-limited case size. The 8008 also generates a sync line to provide a pulse stream to maintain the external system synchronized to the internal CPU clock.

THE BASIC INSTRUCTION CYCLE

The five operation steps of the 8008 master cycle are utilized as follows:

Time Period	S_0	S_1	S_2	Operation
T_1	0	1	0	Lower half of the 8008 address code is sent out on address lines.
T_2	0	0	1	Upper half of the 8008 address code is sent out on address lines (6 bits).
T_3	1	1	0	Data are moved in or out of the 8008 on the address lines.
T_4	1	1	1	Operation instruction is executed internal to the 8008.
T_5	1	0	1	Operation instruction is executed internal to the 8008.

It should be observed that the address sent out of the 8008 processor during the T_2 time period is *not* composed of 8 bits (the normal data word length) but is rather composed of 6 bits. This shortened word enables the 8008 to inform the external processor elements of the operation to be performed by the address just sent out. Since the 8008 is limited to performing all of its communication with any

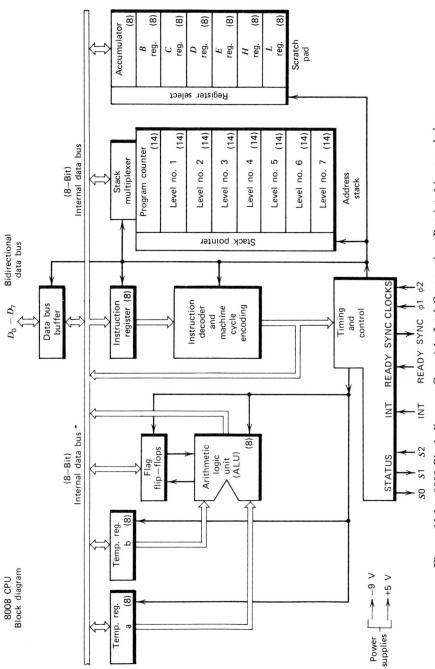

8008 CPU
Block diagram

Figure 11.1. 8008 Block diagram. Copyright Intel Corporation. Reprinted by permission.

125

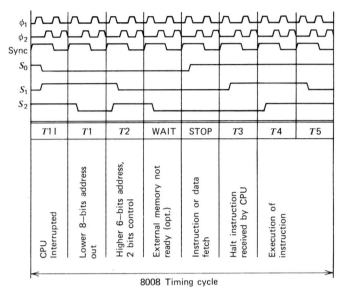

Figure 11.2. Processor timing cycle. Copyright Intel Corporation. Reprinted by permission.

external element over the eight address lines, the two most significant bits (D_7 and D_6) are utilized during the T_2 time period to designate the activity being performed, such as memory read (0,0), memory write (1,1), input port read or output port write (0,1), or memory read data (1,0). In total, four possible activities are conveyed by the D_7,D_6 bits during the T_2 time period that will be acted on during the T_3 time period.

Every instruction cycle starts off in the classical manner with an instruction fetch from the ROM program memory. This instruction fetch occurs during the T_1 and T_2 time periods in accordance with the 14-bit wide address code. The D_7 and D_6 bits instruct the external system that this is an instruction fetch, the 8-bit instruction being read into the 8008 during the T_3 time period where it is stored in the instruction register.

The one byte (8-bit) instruction, which is now residing in the instruction register, is sufficient for the complete execution of an instruction as long as the data to be used in the operation are already located within one of the on-chip registers in the 8008 chip (i.e., the instruction does not require data from memory or from the input/output ports). The execution of the instruction for "on-chip" data is performed during the T_4 and T_5 time periods.

The instruction byte that is held within the instruction register is actually composed of a double 3-bit address and a 2-bit operation code as shown below:

XX	*DDD*	*SSS*
Operation code	Destination address	Source address

THE ON-CHIP REGISTERS

The 8008 chip contains seven (7) on-chip operating 8-bit registers and one dummy memory register that points to the H and L registers used for addressing external memory. The addresses of these on-chip memories are as follows:

Address	Register	
000	*A*	(accumulator)
001	*B*	(auxiliary accumulators)
010	*C*	
011	*D*	(also used for high-order external memory address)
100	*E*	
101	*H*	(also used for low-order external memory address)
110	*L*	
111	*M*	(dummy address that points to *H,L* registers)

The first seven of these are working registers within an 8×7 dynamic on-chip RAM that is automatically refreshed. Data may be moved in or out of these registers by using the appropriate operation code and addressing the working register in the desired source or designation location. For example, if it is desired to move the contents of register B to register A [$(A) \leftarrow (B)$ in shorthand notation], an op-code "11" would be used and the instruction would be written as follows:

11   000   001   where the source of the data is register B (address 001)
$\underbrace{}_{DDD}$ $\underbrace{}_{SSS}$ and the destination of the data is register A (address 000).

If, however, data from external memory (off chip memory) or the input ports are required to perform the desired operation, the appropriate source or destination address location would be filled by the eighth register (M) address, 111. This code automatically directs the 8008 to issue a second master machine cycle (T_1–T_5). During this second master cycle, the contents of the L and H registers are sent out during the T_1 and T_2 time periods and act as the memory-location address byte. The L address contains the full 8-bit word corresponding to the lower half of the program counter address (256 byte locations), while the H address contains only 6 bits of address (64 pages), the uppermost bit positions D_7 and D_6 instructing the external system what form of activity is to take place during the following time period T_3 when data are to be moved.

D_7 and D_6 will set the read/write line to direct the RAM memory to "read" if the 111 address is in the source position in the instruction register, or to "write" if the 111 address is in the destination position in the instruction register.

THE 8008 INSTRUCTION SET

The index register and accumulator-group instructions are shown in Figure 11.3. An explanation of each instruction (copywrite Intel Corporation—reprinted by permission) is provided in the tabular lists following Figure 11.3.

Load Data to Index Registers—One Byte

Data may be loaded into or moved between any of the index registers, or memory registers.

Lr_1r_2 (one cycle— PCI)	11	*DDD*	*SSS*	$(r_1) \leftarrow (r_2)$ Load register r_1 with the content of r_2. The content of r_2 remains unchanged. If *SSS* = *DDD* the instruction is a NOP (no operation).
LrM (two cycles— PCI/PCR)	11	*DDD*	111	$(r) \leftarrow (M)$ Load register r with the content of the memory location addressed by the contents of registers H and L. (*DDD* $\neq$ 111—HALT instruction.)
LMr (two cycles— PCI/PCW)	11	111	*SSS*	$(M) \leftarrow (r)$ Load the memory location addressed by the contents of registers H and L with the content of register r. (*SSS* $\neq$ 111—HALT instruction).

Load Data Immediate—2 Bytes

A byte of data immediately following the instruction may be loaded into the processor or into the memory.

LrI (two cycles— PCI/PCR)	00	*DDD* $\langle B_2 \rangle$	110⌉ ⌋	$(r) \leftarrow \langle B_2 \rangle$ Load byte two of the instruction into register r.

BASIC INSTRUCTION SET

Data and Instruction Formats

Data in the 8008 is stored in the form of 8-bit binary integers. All data transfers to the system data bus will be in the same format.

$$\boxed{D_7\ D_6\ D_5\ D_4\ D_3\ D_2\ D_1\ D_0}$$
DATA WORD

The program instructions may be one, two, or three bytes in length. Multiple byte instructions must be stored in successive words in program memory. The instruction formats then depend on the particular operation executed.

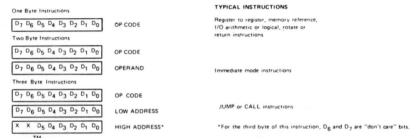

For the MCS-8™ a logic "1" is defined as a high level and a logic "0" is defined as a low level.

Figure 11.3. 8008 Instructions. Copyright Intel Corporation. Reprinted by permission.

INDEX REGISTER INSTRUCTIONS

The load instructions do not affect the flag flip-flops. The increment and decrement instructions affect all flip-flops except the carry.

MNEMONIC	MINIMUM STATES REQUIRED	D_7 D_6	D_5 D_4 D_3	D_2 D_1 D_0	DESCRIPTION OF OPERATION
(1)Lr_1r_2	(5)	1 1	D D D	S S S	Load index register r_1 with the content of index register r_2.
(2)LrM	(8)	1 1	D D D	1 1 1	Load index register r with the content of memory register M.
LMr	(7)	1 1	1 1 1	S S S	Load memory register M with the content of index register r.
(3)Lrl	(8)	0 0 B B	D D D B B B	1 1 0 B B B	Load index register r with data B . . . B.
LMI	(9)	0 0 B B	1 1 1 B B B	1 1 0 B B B	Load memory register M with data B . . . B.
INr	(5)	0 0	D D D	0 0 0	Increment the content of index register r (r ≠ A).
DCr	(5)	0 0	D D D	0 0 1	Decrement the content of index register r (r ≠ A).

ACCUMULATOR GROUP INSTRUCTIONS

The result of the ALU instructions affect all of the flag flip-flops. The rotate instructions affect only the carry flip-flop.

MNEMONIC	MINIMUM STATES REQUIRED	D_7 D_6	D_5 D_4 D_3	D_2 D_1 D_0	DESCRIPTION OF OPERATION
ADr	(5)	1 0	0 0 0	S S S	Add the content of index register r, memory register M, or data
ADM	(8)	1 0	0 0 0	1 1 1	B . . . B to the accumulator. An overflow (carry) sets the carry
ADI	(8)	0 0 B B	0 0 0 B B B	1 0 0 B B B	flip-flop.
ACr	(5)	1 0	0 0 1	S S S	Add the content of index register r, memory register M, or data
ACM	(8)	1 0	0 0 1	1 1 1	B . . . B to the accumulator with carry. An overflow (carry)
ACI	(8)	0 0 B B	0 0 1 B B B	1 0 0 B B B	sets the carry flip-flop.
SUr	(5)	1 0	0 1 0	S S S	Subtract the content of index register r, memory register M, or
SUM	(8)	1 0	0 1 0	1 1 1	data B . . . B from the accumulator. An underflow (borrow)
SUI	(8)	0 0 B B	0 1 0 B B B	1 0 0 B B B	sets the carry flip-flop.
SBr	(5)	1 0	0 1 1	S S S	
SBM	(8)	1 0	0 1 1	1 1 1	Subtract the content of index register r, memory register M, or data
SBI	(8)	0 0 B B	0 1 1 B B B	1 0 0 B B B	data B . . . B from the accumulator with borrow. An underflow (borrow) sets the carry flip-flop.
NDr	(5)	1 0	1 0 0	S S S	Compute the logical AND of the content of index register r,
NDM	(8)	1 0	1 0 0	1 1 1	memory register M, or data B . . . B with the accumulator.
NDI	(8)	0 0 B B	1 0 0 B B B	1 0 0 B B B	
XRr	(5)	1 0	1 0 1	S S S	Compute the EXCLUSIVE OR of the content of index register
XRM	(8)	1 0	1 0 1	1 1 1	r, memory register M, or data B . . . B with the accumulator.
XRI	(8)	0 0 B B	1 0 1 B B B	1 0 0 B B B	
ORr	(5)	1 0	1 1 0	S S S	Compute the INCLUSIVE OR of the content of index register
ORM	(8)	1 0	1 1 0	1 1 1	r, memory register m, or data B . . . B with the accumulator.
ORI	(8)	0 0 B B	1 1 0 B B B	1 0 0 B B B	
CPr	(5)	1 0	1 1 1	S S S	Compare the content of index register r, memory register M,
CPM	(8)	1 0	1 1 1	1 1 1	or data B . . . B with the accumulator. The content of the
CPI	(8)	0 0 B B	1 1 1 B B B	1 0 0 B B B	accumulator is unchanged.
RLC	(5)	0 0	0 0 0	0 1 0	Rotate the content of the accumulator left.
RRC	(5)	0 0	0 0 1	0 1 0	Rotate the content of the accumulator right.
RAL	(5)	0 0	0 1 0	0 1 0	Rotate the content of the accumulator left through the carry.
RAR	(5)	0 0	0 1 1	0 1 0	Rotate the content of the accumulator right through the carry.

Figure 11.3. (continued)

PROGRAM COUNTER AND STACK CONTROL INSTRUCTIONS

(4) JMP	(11)	0 1	X X X	1 0 0	Unconditionally jump to memory address $B_3 \ldots B_3 B_2 \ldots B_2$.
		$B_2\ B_2$	$B_2\ B_2\ B_2$	$B_2\ B_2\ B_2$	
		X X	$B_3\ B_3\ B_3$	$B_3\ B_3\ B_3$	
(5) JFc	(9 or 11)	0 1	0 $C_4\ C_3$	0 0 0	Jump to memory address $B_3 \ldots B_3 B_2 \ldots B_2$ if the condition
		$B_2\ B_2$	$B_2\ B_2\ B_2$	$B_2\ B_2\ B_2$	flip-flop c is false. Otherwise, execute the next instruction in sequence.
		X X	$B_3\ B_3\ B_3$	$B_3\ B_3\ B_3$	
JTc	(9 or 11)	0 1	1 $C_4\ C_3$	0 0 0	Jump to memory address $B_3 \ldots B_3 B_2 \ldots B_2$ if the condition
		$B_2\ B_2$	$B_2\ B_2\ B_2$	$B_2\ B_2\ B_2$	flip-flop c is true. Otherwise, execute the next instruction in sequence.
		X X	$B_3\ B_3\ B_3$	$B_3\ B_3\ B_3$	
CAL	(11)	0 1	X X X	1 1 0	Unconditionally call the subroutine at memory address $B_3 \ldots$
		$B_2\ B_2$	$B_2\ B_2\ B_2$	$B_2\ B_2\ B_2$	$B_3 B_2 \ldots B_2$. Save the current address (up one level in the stack).
		X X	$B_3\ B_3\ B_3$	$B_3\ B_3\ B_3$	
CFc	(9 or 11)	0 1	0 $C_4\ C_3$	0 1 0	Call the subroutine at memory address $B_3 \ldots B_3 B_2 \ldots B_2$ if the
		$B_2\ B_2$	$B_2\ B_2\ B_2$	$B_2\ B_2\ B_2$	condition flip-flop c is false, and save the current address (up one
		X X	$B_3\ B_3\ B_3$	$B_3\ B_3\ B_3$	level in the stack.) Otherwise, execute the next instruction in sequence.
CTc	(9 or 11)	0 1	1 $C_4\ C_3$	0 1 0	Call the subroutine at memory address $B_3 \ldots B_3 B_2 \ldots B_2$ if the
		$B_2\ B_2$	$B_2\ B_2\ B_2$	$B_2\ B_2\ B_2$	condition flip-flop c is true, and save the current address (up one
		X X	$B_3\ B_3\ B_3$	$B_3\ B_3\ B_3$	level in the stack). Otherwise, execute the next instruction in sequence.
RET	(5)	0 0	X X X	1 1 1	Unconditionally return (down one level in the stack).
RFc	(3 or 5)	0 0	0 $C_4\ C_3$	0 1 1	Return (down one level in the stack) if the condition flip-flop c is false. Otherwise, execute the next instruction in sequence.
RTc	(3 or 5)	0 0	1 $C_4\ C_3$	0 1 1	Return (down one level in the stack) if the condition flip-flop c is true. Otherwise, execute the next instruction in sequence.
RST	(5)	0 0	A A A	1 0 1	Call the subroutine at memory address AAA000 (up one level in the stack).

INPUT/OUTPUT INSTRUCTIONS

INP	(8)	0 1	0 0 M	M M 1	Read the content of the selected input port (MMM) into the accumulator.
OUT	(6)	0 1	R R M	M M 1	Write the content of the accumulator into the selected output port (RRMMM, RR ≠ 00).

MACHINE INSTRUCTION

HLT	(4)	0 0	0 0 0	0 0 X	Enter the STOPPED state and remain there until interrupted.
HLT	(4)	1 1	1 1 1	1 1 1	Enter the STOPPED state and remain there until interrupted.

NOTES:
(1) SSS = Source Index Register ⎤ These registers, r_i, are designated A(accumulator–000),
     DDD = Destination Index Register ⎦ B(001), C(010), D(011), E(100), H(101), L(110).
(2) Memory registers are addressed by the contents of registers H & L.
(3) Additional bytes of instruction are designated by BBBBBBBB.
(4) X = "Don't Care".
(5) Flag flip-flops are defined by $C_4 C_3$: carry (00-overflow or underflow), zero (01-result is zero), sign (10-MSB of result is "1"), parity (11-parity is even).

intel

Figure 11.3. (continued)

LMI   00   111   110 ⎤ $(M) \leftarrow \langle B_2 \rangle$ Load byte two of the instruction into
(three cycles— $\langle B_2 \rangle$ ⎦   the memory location addressed by the contents
PCI/PCR/PCW)   of registers *H* and *L*.

Increment Index Register—One Byte

INr   00   *DDD*   000 $(r) \leftarrow (r) + 1$. The content of register *r* is in-
(one cycle—   creased by one. All of the condition flip-flops
PCI)   except carry are affected by the result. Note that
                                                                   $DDD \neq 000$ (HALT instruction) and $DDD \neq$
                                                                   111 (content of memory may not be increased).

Decrement Index Register—One Byte

DCr   00   *DDD*   001 $(r) \leftarrow (r) - 1$. The content of register *r* is de-
(one cycle—   creased by one. All of the condition flip-flops
PCI)   except carry are affected by the result. Note that
                                                                   $DDD \neq 000$ (HALT instruction) and $DDD \neq$
                                                                   111 (content of memory may not be decremented).

Accumulator-group Instructions

Operations are performed and the status flip-flops, *C, Z, S, P*, are set based on the
result of the operation. Logical operations (*NDr, XRr, ORr*) set the carry flip-flop
to zero. Rotate operations affect only the carry flip-flop. Two's complement subtraction is used.

ALU Index-register Instructions—One Byte (One Cycle—PCI)

Index-register operations are carried out between the accumulator and the content
of one of the index registers (*SSS* = 000 through *SSS* = 110). The previous content of register *SSS* is unchanged by the operation.

ADr   10   000   *SSS* $(A) \leftarrow (A) + (r)$ Add the content of register *r* to
                                                                   the content of register *A* and place the result into
                                                                   register *A*.

ACr   10   001   *SSS* $(A) \leftarrow (A) + (r) + (carry)$ Add the content of
                                                                   register *r* and the contents of the carry flip-flop
                                                                   to the content of the *A* register and place the result
                                                                   into register *A*.

SUr   10   010   *SSS* $(A) \leftarrow (A) - (r)$ Subtract the content of register *r*
                                                                   from the content of register *A* and place the result
                                                                   into register *A*. Two's complement subtraction is
                                                                   used.

SBr	10	011	SSS	$(A) \leftarrow (A) - (r) -$ (borrow) Subtract the content of register r and the content of the carry flip-flop from the content of register A and place the result into register A.
NDr	10	100	SSS	$(A) \leftarrow (A) \wedge (r)$ Place the logical product of the register A and register r into register A.
XRr	10	101	SSS	$(A) \leftarrow (A) \veebar (r)$ Place the "exclusive – OR" of the content of register A and register r into register A.
ORr	10	110	SSS	$(A) \leftarrow (A) \vee (r)$ Place the "inclusive – OR" of the content of register A and register r into register A.
CPr	10	111	SSS	$(A) - (r)$ Compare the content of register A with the content of register r. The content of register r remains unchanged. The flag flip-flops are set by the result of the subtraction.

Equality $(A = r)$ is indicated by the zero flip-flop set to "1." Less than $(A < r)$ is indicated by the carry flip-flop, set to "1."

ALU Operations with Memory—One Byte (Two Cycles—PCI/PCR)

Arithmetic and logical operations are carried out between the accumulator and the byte of data addressed by the contents of registers H and L.

ADM	10	000	111	$(A) \leftarrow (A) + (M)$ **ADD**
ACM	10	001	111	$(A) \leftarrow (A) + (M) +$ (carry) **ADD** with carry
SUM	10	010	111	$(A) \leftarrow (A) - (M)$ **SUBTRACT**
SBM	10	011	111	$(A) \leftarrow (A) - (M) -$ (borrow) **SUBTRACT** with borrow
NDM	10	100	111	$(A) \leftarrow (A) \wedge (M)$ **Logical AND**
XRM	10	101	111	$(A) \leftarrow (A) \veebar (M)$ **Exclusive OR**
ORM	10	110	111	$(A) \leftarrow (A) \vee (M)$ **Inclusive OR**
CPM	10	111	111	$(A) - (M)$ **COMPARE**

ALU Immediate Instructions—Two Bytes (Two Cycles—PCI/PCR)

Arithmetic and logical operations are carried out between the accumulator and the byte of data immediately following the instruction.

ADI	00	000	100	$(A) \leftarrow (A) + \langle B_2 \rangle$
		$\langle B_2 \rangle$		**ADD**
ACI	00	001	100	$(A) \leftarrow (A) + \langle B_2 \rangle +$ (carry)
		$\langle B_2 \rangle$		**ADD** with carry
SUI	00	010	100	$(A) \leftarrow (A) - \langle B_2 \rangle$
		$\langle B_2 \rangle$		**SUBTRACT**
SBI	00	011	100	$(A) \leftarrow (A) - \langle B_2 \rangle -$ (borrow)
		$\langle B_2 \rangle$		**SUBTRACT** with borrow

NDI	00	100	100⌉	$(A) \leftarrow (A) \wedge \langle B_2 \rangle$
		$\langle B_2 \rangle$	⌋	logical AND
XRI	00	101	100⌉	$(A) \leftarrow (A) \not\vee \langle B_2 \rangle$
		$\langle B_2 \rangle$	⌋	exclusive OR
ORI	00	110	100⌉	$(A) \leftarrow (A) \vee \langle B_2 \rangle$
		$\langle B_2 \rangle$	⌋	inclusive OR
CPI	00	111	100⌉	$(A) - \langle B_2 \rangle$
		$\langle B_2 \rangle$	⌋	COMPARE

Rotate Instructions—One Byte (One Cycle—PCI)

The accumulator content (register A) may be rotated (shifted) either right or left, around the carry bit or through the carry bit. Only the carry flip-flop is affected by these instructions; the other flags are unchanged.

RLC   00   000   010 $A_{m+1} \leftarrow A_m, A_0 \leftarrow A_7, (\text{carry}) \leftarrow A_7$
Rotate the content of register A left one bit. Rotate A_7 into A_0 and into the carry flip-flop.

RRC   00   001   010 $A_m \leftarrow A_{m+1}, A_7 \leftarrow A_0, (\text{carry}) \leftarrow A_0$
Rotate the content of register A right one bit. Rotate A_0 into A_7 and into the carry flip-flop.

RAL   00   010   010 $A_{m+1} \leftarrow A_m, A_0 \leftarrow (\text{carry}), (\text{carry}) \leftarrow A_7$
Rotate the content of register A left one bit. Rotate the content of the carry flip-flop into A_0. Rotate A_7 into the carry flip-flop.

RAR   00   011   010 $A_m \leftarrow A_{m+1}, A_7 \leftarrow (\text{carry}), (\text{carry}) \leftarrow A_0$
Rotate the content of register A right one bit. Rotate the content of the carry flip-flop into A_7. Rotate A_0 into the carry flip-flop.

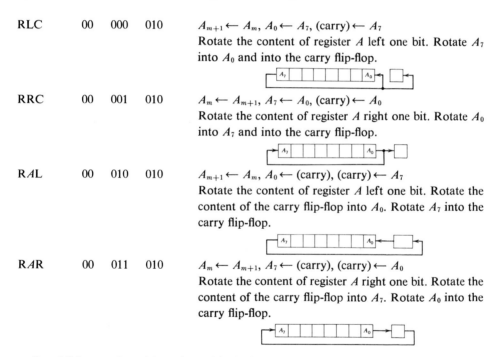

In addition to the arithmetic and logic functions of the 8008 processor, the decision making or "smart terminal" portion of the microprocessor is found within the jump/call/return subroutines of the device. Without these instructions, the microprocessor is little more than a robot capable of performing simple mathematical functions. Armed with the jump, call, and return instructions, the machine can alter its program dependent on outside conditions or data results. In short, the processor becomes a smart terminal.

The **jump unconditionally** instruction dictates, under program control, a departure from the normal program sequence to the program address indicated by the second and third bytes of the three byte instruction. This instruction can also be used to insert within an existing program a subprogram which was written *out of* *sequence*. The two **jump if** instructions are true decision instructions, altering the program sequence *if certain conditions exist* but continuing on the normal program if the conditions do not exist. These two jump conditions are controlled by any one of four flag flip-flops.

The jump instruction does not involve any data, but merely reassigns the program counter to a new number.

The call instruction is similar to the jump instruction except that the call instruction is used when a **return** to the present program counter address is desired. Like the jump instruction, the **call unconditional** instruction dictates a departure from the normal program sequence to the program address indicated by the second and third bytes of the three-byte instruction. This instruction can also be used to insert a subroutine which was written out of sequence into an existing program. The two **call if** instructions are true decision instructions, altering the program sequence if certain conditions exist but continuing on the normal program if the conditions do not exist. These instructions are ideal when short modifications of control functions are desired within the context of a main program sequence depending on the state of any one of four flag flip-flops. Up to seven call subroutines may be utilized (nested) at a time.

The **return** instruction is utilized to exit from a call routine. When the call instruction is initiated, the existing program counter location is "saved" (or pushed down) in the program counter memory stack, the new program counter located at the top of the stack. When the return instruction is initiated, the stack is "popped up" one level exposing the last program counter location previous to the call instruction. The return instruction, like the jump and call instructions, can be either unconditional or conditioned upon the state of one of the four flag flip-flops.

The Flag Flip-flops

The four flag flip-flops controlling the **jump if**, **call if**, and **return if** instructions are the C (carry), Z (zero), S (sign), and P (parity) flip-flops. Within the operation instructions they are referenced by the designator C_4C_3 as follows:

C_4C_3	Flip-flop Referenced
00	carry (C)
01	zero (Z)
10	sign (S)
11	parity (P)

The flags are usually operationally set by the normal computer process. The flag flip-flops are set as follows:

C The carry (or borrow) flip-flop generally denotes overflow (the operation of the accumulator results in a word exceeding the 8-bit byte length, and a "1" overflows into the ninth bit) or underflow (the operation of the accumulator results in a number less than one bit but greater than zero resulting in a "1" occurring (underflowing) to the right of the "0" bit). In either case the carry (C) flip-flop (acting as an extra bit on either end of the accumulator) will be set to 1. If there is no overflow/underflow, the C flag remains set to 0.

During the rotate-left/rotate-right instructions, the C flag will be set to the value of the bit rotated into it.

During logic operations the C flag is always set to 0.

During compare operations the C flag is set to 1 if $[A] < [R]$.

Z The zero flag flip-flop is set to 1 any time an operation is performed that results in the operation equaling zero.

The Z flag is not changed during rotation operations.

S The sign flag flip-flop is set by the D_7 accumulator bit. If D_7 is 1, the S flag flip-flop is set to 1.

The sign flag is not set during rotation operations.

P The parity flag flip-flop is set to 1 if there is an even number of 1 bits in the accumulator.

The parity flag is not set during rotation operations.

A summary of the microprocessor operations effects on the four flag flip-flops is illustrated in Figure 11.4. The detailed explanations of each of the program counter and stack-control instructions are as follows.

Jump Instructions—3 Bytes (Three Cycles—PCI/PCR/PCR)

Normal flow of the microprogram may be altered by jumping to an address specified by bytes two and three of an instruction.

| JMP (jump unconditionally) | 01 | XXX $\langle B_2 \rangle$ $\langle B_3 \rangle$ | 100 | $(P) \leftarrow \langle B_3 \rangle \langle B_2 \rangle$. Jump unconditionally to the instruction located in memory location addressed by bytes 2 and 3. |
| JFc (jump if condition false) | 01 | $0C_4C_3$ $\langle B_2 \rangle$ $\langle B_3 \rangle$ | 000 | If $(c) = 0$, $(P) \leftarrow \langle B_3 \rangle \langle B_2 \rangle$. Otherwise, $(P) = (P) + 3$. If the content of flip-flop c is zero, then jump to the instruction located in memory location $\langle B_3 \rangle \langle B_2 \rangle$; otherwise, execute the next instruction in sequence. |

FLAG BITS

	C	Z	S	P	
Increment/Decrement	—	x	x	x	$C = 1$ for overflow and underflow
Rotation L/R	x	—	—	—	$Z = 1$ for zero
Accumulator operations					$S = 1$, $D_7 = 1$
ADD, SUB, ADC, SBB (ADD, SUB, ADC, SBB) DATA	x	x	x	x	$P = 1$, even #1's
Logic operations AND, ORA, XRA, (AND, ORA, XRA) DATA	$x = 0$	x	x	x	$x \equiv$ Flag bits set by normal operations
CMP, (CMP) DATA $[A - r]$	$A < R$	$A = R$	x	x	

Figure 11.4. Summary of flag-bits.

JTc   01 $1C_4C_3$ 000⎤   If $(c) = 1$, $(P) \leftarrow \langle B_3 \rangle \langle B_2 \rangle$. Otherwise,
(jump if condition true) $\langle B_2 \rangle$ │ $(P) = (P) + 3$. If the content of flip-flop c
                                          $\langle B_3 \rangle$ ⎦   is one, then jump to the instruction
                                                                              located in memory location $\langle B_3 \rangle \langle B_2 \rangle$;
                                                                              otherwise, execute the next instruction in
                                                                              sequence.

Call Instructions—Three Bytes (Three Cycles—PCI/PCR/PCR)

Subroutines may be called and nested up to seven levels.

CAL   01 XXX 110⎤   (Stack) $\leftarrow$ (P) $\leftarrow$ $\langle B_3 \rangle \langle B_2 \rangle$. Shift the
(call subroutine uncon- $\langle B_2 \rangle$ │   content of P to the pushdown stack. Jump
ditionally) $\langle B_3 \rangle$ ⎦   unconditionally to the instruction located
                                                                              in memory location addressed by bytes
                                                                              2 and 3.

CFc   01 $0C_4C_3$ 010⎤   If $(c) = 0$, (Stack) $\leftarrow$ (P) (P) $\leftarrow$ $\langle B_3 \rangle$
(call subroutine if con- $\langle B_2 \rangle$ │ $\langle B_2 \rangle$. Otherwise, $(P) = (P) + 3$. If the
dition false) $\langle B_3 \rangle$ ⎦   content of flip-flop c is zero, then shift
                                                                              contents of P to the pushdown stack and
                                                                              jump to the instruction located in memory
                                                                              location $\langle B_3 \rangle \langle B_2 \rangle$; otherwise execute the
                                                                              next instruction in sequence.

CTc   01 $1C_4C_3$ 010⎤   If $(c) = 1$, (Stack) $\leftarrow$ (P) (P) $\leftarrow$ $\langle B_3 \rangle$
(call subroutine if con- $\langle B_2 \rangle$ │ $\langle B_2 \rangle$. Otherwise, $(P) = (P) + 3$. If the
dition true) $\langle B_3 \rangle$ ⎦   content of flip-flop c is one, then shift
                                                                              contents of P to the pushdown stack and
                                                                              jump to the instruction located in memory
                                                                              location $\langle B_3 \rangle \langle B_2 \rangle$; otherwise execute the
                                                                              next instruction in sequence.

In the above JUMP and CALL instructions $\langle B_2 \rangle$ contains the least significant
half of the address (L) and $\langle B_3 \rangle$ contains the most significant half of the address (H).
Note that D_7 and D_6 of $\langle B_3 \rangle$ are "don't care" bits since the CPU uses 14 bits of
address.

Return Instructions—One Byte (One Cycle—PCI)

A return instruction may be used to exit from a subroutine; the stack is popped up
one level at a time.

RET   00 XXX 111 $(P) \leftarrow$ (Stack). Return to the instruction in the
                                          memory location addressed by the last value shifted
                                          into the pushdown stack. The stack pops up one
                                          level.

RFc (return condition false)	00	0C_4C_3	011	If $(c) = 0, (P) \leftarrow$ (Stack). Otherwise, $(P) = (P) + 1$. If the content of flip-flop c is zero, then return to the instruction in the memory location addressed by the last value inserted in the pushdown stack. The stack pops up one level. Otherwise, execute the next instruction in sequence.
RTc (return condition true)	00	1C_4C_3	011	If $(c) = 1, (P) \leftarrow$ (Stack). Otherwise, $(P) = (P) + 1$. If the content of flip-flop c is one, then return to the instruction in the memory location addressed by the last value inserted in the pushdown stack. The stack pops up one level. Otherwise, execute the next instruction in sequence.

Eight input devices and 24 output devices may be referenced by the Intel 8008 microprocessor. These two instructions are self-explanatory and are detailed as follows.

Input/Output Instructions—1 Byte (Two Cycles—PCI/PCC)

Eight input devices may be referenced by the input instruction.

INP	01	00M	MM1	$(A) \leftarrow$ (input data lines). The content of register A is made available to external equipment at state T_1 of the PCC cycle. The content of the instruction register is made available to external equipment at state T_2 of the PCC cycle. New data for the accumulator are loaded at T_3 of the PCC cycle. MMM denotes input-device number. The content of the condition flip-flops, S, Z, P, C, is output on D_0, D_1, D_2, D_3, respectively at T_4 on the PCC cycle.

Twenty-four output devices may be referenced by the output instruction.

OUT	01	RRM	MM1	(Output data lines) $\leftarrow (A)$. The content of register A is made available to external equipment at state T_1, and the content of the instruction register is made available to external equipment at state T_2 of the PCC cycle. $RRMMM$ denotes output device number ($RR \neq 00$).

The final two instructions in the Intel 8008 programming stable are the RESTART instruction and the HALT instruction, the latter is obvious and is not elaborated on. RESTART instruction, however, is more unique.

Once the machine is halted for any reason, the machine can be restarted in any one of eight subroutines. Thus, a capability of eight major, independent "dedicated" programs could be located in a single microprocessor, each being on immediate call by the operator at the conclusion of a previous program. The detailed explanation of the two machine operations are as follows.

HALT Instruction—1 Byte (One Cycle—PCI)

HLT	00	000	00X	On receipt of the HALT instruction the activity of

| | | or | | the processor is immediately suspended in the |
| | 11 | 111 | 111 | STOPPED state. The content of all registers and |

memory remains unchanged. The P counter has been updated and the internal dynamic memories continue to be refreshed.

RESTART Instruction—1 Byte (One Cycle—PCI)

The RESTART instruction acts as a one byte call on eight specified locations on page 0, the first 256 instruction words.

RST	00	AAA	101	$(\text{Stack}) \leftarrow (P), (P) \leftarrow (000000\ 00AAA000)$. Shift the

contents of P to the pushdown stack. The content, AAA, of the instruction register is shifted into bits 3–5 of the P counter. All other bits of the P counter are set to zero. As a one-word "call," 8-byte subroutines may be accessed in the lower 64 words of memory.

PROBLEMS

1. Determine the binary codes for the following machine operations:
   a. register $A \rightarrow$ register L;
   b. data (5_{10}) loaded into register B;
   c. register $A \rightarrow$ output register 15;
   d. register L incremented by 1;
   e. register A compared to 25_{10}.

2. Write the binary program to perform the following machine operations:
   a. Add register B to register C and output the result to output register 15.
   b. Input data from input register 2 and store the data in memory location 42_{10}.
   c. Input data from input register 5 and store the data in register C.
   d. Compare register A to 105_{10}. If $[A] = 105_{10}$, add the contents of register A to register C and output the total to output register 10. If $[A] \neq 105_{10}$, output the contents of register A to ouput register 15.

3. Given the following machine-operation flowcharts, write the machine level programs:

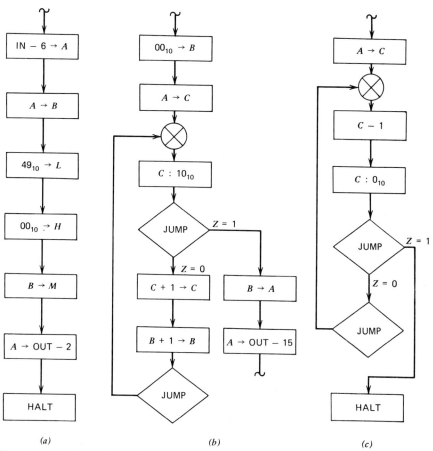

(a)   (b)   (c)

4. Develop the logic flowchart, machine-operation flowchart, and machine-level program for an electronic security lock if the combination required is your birthdate *and* your age taken one at a time. Input your birthdate into input register 00 and your age into input register 01. The lock release code FF will be outputted to output register 20.

5. Develop the flowcharts (logic and machine operation) to perform the logical operation:

$$A \cdot (B + C \cdot D) \vee (D + C \cdot A) \cdot B = F$$

Input $A \rightarrow$ register 0, $B \rightarrow$ register 1, $C \rightarrow$ register 2, and $D \rightarrow$ register 3. Output the result $F \rightarrow$ register 18.

CHAPTER 12

The Hexadecimal Operation Codes

THE COMPUTER NUMBER SYSTEMS

The instruction codes developed for the microprocessor are 8-bit binary words that appear identical in format to 8-bit data words or 8-bit address words. In short, there is no perceptible difference between the three forms of binary words except as defined by the programmer and used in the microprocessor. All three forms of word are composed of 8 bits of binary data. However, while this binary-data format is easily recognized by the microprocessor system, it is tedious for the programmer to use. It is imperative, therefore, to find a higher-order coding format that can be compatible with both the processor and the programmer.

The Octal Number System

Two higher-order coding formats exist. The first format is called the octal code. This code is used to designate the decimal value of the binary word taken three bits at a time. Three binary bits (a_2, a_1, a_0) can be used to designate a total of eight possible digits (states) and thus the term *octal* is used. A listing of the relationship between the decimal, octal, and binary based codes is shown in Table 12.1.

TABLE 12.1. Decimal—Octal—Binary Code Conversions

Decimal	Octal	Binary
0	0	000
1	1	001
2	2	010
3	3	011
4	4	100
5	5	101
6	6	110
7	7	111
8(0)	0	000
9(1)	1	001

The number 377 in octal is translated into a binary code by a combination of three 3-bit binary bytes yielding 011,111,111, which can be read as an 8-bit binary number 11111111. To code the 8-bit binary number into the input register, the operator is required only to push the number 377 on the octal-coded keyboard. It can be seen from Table 12.1 that a major limitation in decimal-to-octal number conversion exists. The maximum counting range of the octal system is a function of 7 (i.e., 77,777,7777, etc.) rather than the decimal 9 (i.e., 99,999,9999, etc.). This means that the full use of the decimal-to-binary conversion cannot really be utilized. In addition, the programmer may have as much difficulty thinking in the octal system as he has in the binary system.

The Hexadecimal Number System

The second number system used as a higher-order coding format is the hexadecimal system. In contrast to the 3-bit binary word translated by the octal-code format, the hexadecimal code is used to designate the decimal value of a 4-bit binary word. Thus, the hexadecimal code covers all 10 decimal digits with six locations left over. These six locations are given either letter names or symbol designations. Table 12.2 illustrates the hex–decimal–binary code conversion system.

A word written in the hexadecimal code is expressed almost identically in the decimal system. The code 4B is translated into the binary code by two 4-bit binary bytes as follows: 0100, 1011. This input would be read by the processor from a hex-coded keyboard as an input binary word containing 8 bits equal to 01001011.

The hexadecimal code is obviously the easiest system code for the programmer to use since he can virtually think in decimal logic and, therefore, does not have to reorient his thinking when communicating with the machine. In addition, the hexadecimal code utilizes completely the number capacity of the decimal system and vice versa.

TABLE 12.2. Decimal–Hexadecimal–Binary Conversion Codes

Decimal	Hexadecimal	Binary
0	0	0000
1	1	0001
2	2	0010
3	3	0011
4	4	0100
5	5	0101
6	6	0110
7	7	0111
8	8	1000
9	9	1001
10	A	1010
11	B	1011
12	C	1100
13	D	1101
14	E	1110
15	F	1111

THE HEXADECIMAL OPERATION CODE SET

The instruction code set for most microprocessors is presented in a binary coding format in order to generalize each of the codes with respect to the source and destination of data transfer as well as flag flip-flop codes and input/output register locations. However, the use of binary coding is extremely awkward for the programming function since the programmer does not "think" in terms of binary numbers but is trained to work in a decimal format. The limitations of both the binary and octal formats have previously been detailed with respect to the man-machine interface, concluding that the hexadecimal format is the ideal programming code for this interface.

Binary-to-hexadecimal Code Conversion

The transformation of an operational instruction from the 8-bit operation code format to a hexadecimal op-code format is performed as follows:

1. Determine the binary operation code to be transformed. This op-code contains 8 bits (or more) of binary information as shown below:

$$D_7 \mid D_6 \mid D_5 \mid D_4 \mid D_3 \mid D_2 \mid D_1 \mid D_0 \quad \text{op-code}$$

2. Separate the binary 8-bit op-code into two (or more) binary 4-bit codes, D_7–D_4 forming the first code, whereas D_3–D_0 forms the second code.

$$\underbrace{D_7 \mid D_6 \mid D_5 \mid D_4}_{\text{code 1}} \qquad \underbrace{D_3 \mid D_2 \mid D_1 \mid D_0}_{\text{code 2}}$$

3. Transform each of the two (or more) 4-bit codes into an equivalent hexadecimal code. The two-digit hexadecimal code is of identical value to the original 8-bit binary code.

$$H_1, \quad H_2 \qquad \text{hexadecimal op-code}$$

As an example of this procedure, consider the transformation of the operation code for moving the contents of register C into register L. Referring to the instruction codes in Chapter 11, this binary op-code would be:

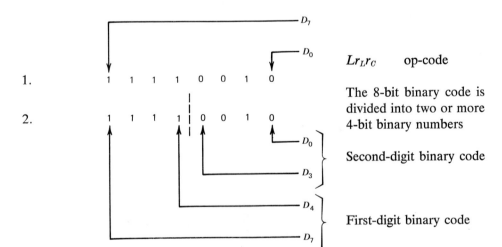

3. F, 2   The two 4-bit binary numbers are converted to their hexadecimal-code equivalent from Table 12.2.

This transformation from binary to hexadecimal coding, while extremely simple, can become very laborious if required to be continuously performed during the programming activity. To alleviate this problem, a matrix of the hexadecimal op-codes for the Intel 8008 microprocessor has been developed and is shown in Figure 12.1.

THE HEXADECIMAL INTEL 8008 OPERATION CODE MATRIX

The Intel 8008 Hexadecimal Operation Code Matrix contains the complete programming code capability for the 8008 microprocessor. The first digit of the hexadecimal operation code (bits (D_7–D_4 of the binary code) is shown on the left-hand side of the matrix, whereas the second hexadecimal digit (bits D_3–D_0 of the binary code) is shown across the top of the matrix. The equivalent binary codes are shown on the right-side and bottom of the matrix for convenience.

To utilize the matrix to its fullest capability, the programmer should become familiar with the location of the various 8008 microprocessor operations within the matrix. To facilitate this familiarization, the matrix can be dissected into its overall op-code blocks in order to illustrate their general locations prior to utilizing the matrix for program generation.

The matrix can be broken into three main horizontal blocks (I,II,III) and two vertical blocks (A,B) as shown in Figure 12.2. In general, the vertical A section appears as a mirror image of the vertical B section. Section A contains the portion of the specific instruction codes associated with register *A, C, E,* and *L,* the condition flip-flops *C* and *S* and the logic operations **ADD, SUB, AND,** and **ORA.** Section B completes the operation code set providing the codes for registers *B, D, H, M,* the condition flip-flops *Z* and *P* and the **ADC, SBB, XRA,** and **CMP** logic operations. For example, in section A the four operation codes for DIRECT DATA manipulation (ADD, SUB, AND, ORA) are listed as 04, 14, 24, and 34 respectively. In section B the remaining four operation codes for DIRECT DATA manipulation (ADC, SBB, XRA, CMP) are listed as 0*C*, 1*C*, 2*C*, and 3*C*, respectively. This "mirror image" can be verified for virtually all of the operation codes listed in the matrix.

The horizontal section I of the op-code matrix contains the codes for **INCREMENTING** and **DECREMENTING** the on-chip registers, **ROTATE LEFT** and **ROTATE RIGHT** codes, all **RETURN** codes, all **DIRECT DATA** manipulating codes, all **RESET** codes and the primary **HALT** codes. Section II of the matrix contains the **INPUT** and **OUTPUT** register codes, all **JUMP** codes and all **CALL** codes. Section III contains all of the **ACCUMULATOR** group instruction and REGISTER instruction codes.

The microprocessor operation codes can be 1 byte in length, 2 bytes in length (DIRECT DATA manipulation codes) or 3 bytes in length (JUMP or CALL codes). The number of bytes which an instruction requires is coded within the matrix by a bracket number ([2], [3]) if the instruction requires more than a single byte code. At the bottom of the matrix sheet the instruction code format corre-

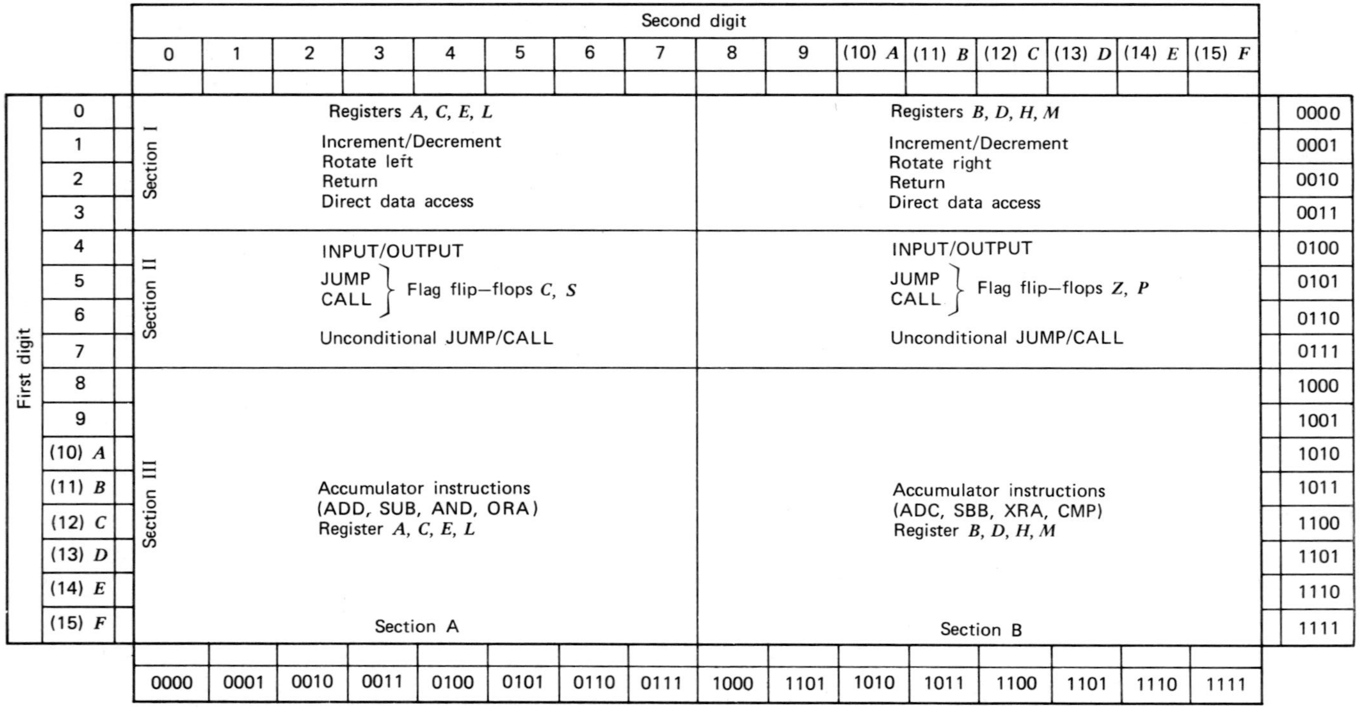

Second digit

First digit	0	1	2	3	4	5	6	7	8	9	(10) A	(11) B	(12) C	(13) D	(14) E	(15) F
	0000	0001	0010	0011	0100	0101	0110	0111	1000	1001	1010	1011	1100	1101	1110	1111
0 0000																
1 0001																
2 0010																
3 0011																
4 0100																
5 0101																
6 0110																
7 0111																
8 1000																
9 1001																
(10) A 1010																
(11) B 1011																
(12) C 1100																
(13) D 1101																
(14) E 1110																
(15) F 1111																

Section I (Section A)
- Registers A, C, E, L
- Increment/Decrement
- Rotate left
- Return
- Direct data access

Section I (Section B)
- Registers B, D, H, M
- Increment/Decrement
- Rotate right
- Return
- Direct data access

Section II (Section A)
- INPUT/OUTPUT
- JUMP / CALL $\}$ Flag flip-flops C, S
- Unconditional JUMP/CALL

Section II (Section B)
- INPUT/OUTPUT
- JUMP / CALL $\}$ Flag flip-flops Z, P
- Unconditional JUMP/CALL

Section III (Section A)
- Accumulator instructions (ADD, SUB, AND, ORA)
- Register A, C, E, L

Section III (Section B)
- Accumulator instructions (ADC, SBB, XRA, CMP)
- Register B, D, H, M

Figure 12.2. Operation-code locator—hexadecimal 8008 operation-code matrix.

144

sponding to the three possible byte lengths is detailed. In addition, the flag flip-flop designations are summarized together with the normal "set" (1) condition requirement.

USE OF THE INTEL 8008 OPERATION CODE MATRIX

The programmer's first task in creating a computer program is the development of the logic and machine operation flow diagrams. Once the detailed machine operation flow diagram has been developed, the programmer enters the matrix to locate each of the operations called for in the flow diagram. When the desired operation is located, the hexadecimal code is determined by first reading the horizontal (left-hand column) code (first digit) and then reading the vertical (top) code (second digit). The two hexadecimal digits then form the desired hexadecimal machine instruction code.

The full capability of the Intel 8008 Hexadecimal Operation Code Matrix can be best demonstrated by the use of several examples.

EXAMPLE 1.

Problem

It is desired to move the data located in register C to the L register. Determine the operation code for this data transfer.

Solution

a. Data transfer between registers would occur in section III, while the L register operations are located in section A. Thus, this operation would be located in the III-A block.

b. Movement occurs from the generalized register to the specific register yielding the resulting location at F (left-hand digit) 2 (top digit).

c. The hexadecimal operation code for moving data from register C to register L is $F2$.

EXAMPLE 2.

Problem

Register D is used as a counter. It is desired to decrement D by 1. Determine the hexadecimal code for this operation.

Solution

a. Incrementing/decrementing operations are found in section I, whereas operations involving the D register are found in section B. This operation is located in section I-B.

b. The operation decrementing register D is located at 1 (left-hand digit) 9 (top digit).

c. The hexadecimal code for decrementing register D is 19.

EXAMPLE 3.

Problem

Transfer from the main program to location 01, 3F (01 being the higher order address while 3F is the lower-order address) if two numbers are identical. We wish to return to the main program at a later time.

Solution

a. During a comparison operation the Z flag is set to 1 if there **is** equality. In addition, since the transfer includes a return to the main program at a later time, the operation would be a CALL rather than the JUMP that does not nest the main program address (PCC). The desired operation is: CALL if $Z = 1$.

b. The CALL operations are located in section II and the Z flag operations, in section B. The operation is located in section II-B.

c. The operation CALL if $Z = 1$ is located at 6 (first digit) A (second digit), yielding an operation code 6A. However, the matrix denotes that the CALL operation requires a 3-byte code (shown by the small 3 in parenthesis following the CALL designation). The first byte of the three byte code is the instruction code (6,A), the second byte the lower-order (L) address location being transferred to (3,F), and the third byte the higher-order (H) address location being transferred to (0,1).

d. The hexadecimal code for the CALL if $Z = 1$ operation with transfer to location address 01, 3F is

$$\begin{array}{ll} 6A & \text{operation code for CALL if } Z = 1 \\ 3F & L \text{ address} \\ 01 & H \text{ address} \end{array}$$

THE PROGRAMMING SHEET

The use of a formalized programming sheet can greatly enhance the facility of writing computer programs. This is especially important in machine-language programming since all JUMP and CALL operations must be directly followed by the 2-byte address of the new location of the next step. Unless there is a means of easily and accurately determining this new location, there will be many blanks left in the program that must be *picked up* before the program is usable.

The program sheet illustrated in Figure 12.3 has been used for programming with favorable results. The programming sheet consists of space for constructing the logic and machine operation flow diagrams with appropriate assertion boxes, a hexadecimal operation code column and the 2-byte ROM code address columns. Since virtually all machine-coded programs are destined to be incorporated into ROM memory by either PROM programming or by supplier mask programming, maintaining accurate ROM location addresses is essential.

The use of a 16 address-code sheet has been adopted since the size of the flow diagram dictates the consumption of the greatest amount of the area on the sheet. The memory page (256 bytes of memory) requires 16 pages of coding sheets.

The remaining information called for on the programming sheet is basic bookkeeping data to permit the programmer to keep track of each program he has

Title: ______________________________ Page: ___ of _____

Programmer: ______________________________ Date: / /

Revision: ________

Logic flow	Machine–operation flow	OP. Code	ROM Addr. H	ROM Addr. L
				0
				1
				2
				3
				4
				5
				6
				7
				8
				9
				A
				B
				C
				D
				E
				F

Figure 12.3. Programming code sheet.

written. This data includes the program title, the date he worked on it, the page number, and finally the revision number of the program in case multiple revisions are made.

The use of the sheet is directly tied to the programmer's capability to reduce the program into a machine-operation flowchart prior to writing the actual program. The steps in using the programming sheet are as follows:

1. The two-byte ROM location code is inserted into the coding-address location to establish the program location within ROM memory.

2. The logic level flowchart is developed in the designated portion of the form.

3. The machine-level operation flowchart is developed in the designated portion of the form. This flowchart must have a box corresponding to each machine function desired.

4. Once a complete flow diagram has been developed, the hexadecimal operation code corresponding to each box of the machine-level operation flowchart is filled in utilizing the Intel 8008 Hexadecimal Operation Code Matrix.

5. The ROM memory is directly programmed from the appropriate three columns (ROM location address (H and L) and Hexadecimal Operation Code).

An example of this form of programming is shown in Figure 12.4.

EXAMPLE: Solve the equation $A = B + C + D$ and output the result to output register 10. (Assume B = data in register B, C = data in register C and D = data in register D.) ROM location is 00, 40.

Title: _____Solution To $A = B + C + D$_____   Page: _1_ of _1_

Programmer: _____W. F. Leahy_____   Date: _/ /_

Revision: _____

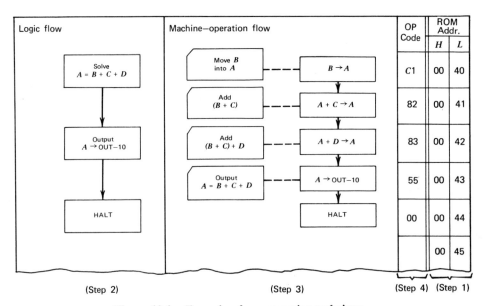

Figure 12.4. Example of programming technique.

PROBLEMS

1. Machine program the equation $x = 3y + 7z - k$. Input y, z, and k into register 0, 1, and 2, respectively. Output x to register 12. Show the logic and machine operation flowcharts.

2. Using the Intel 8008 Operation Code Matrix, determine the hexadecimal code for the following operations:

   1. $A \rightarrow L$
   2. $05 \rightarrow B$
   3. Jump if $z = 1$ to $H = 03$, $L = 4F$
   4. $B \rightarrow A$
   5. $A \rightarrow$ out $- 13$
   6. In $- 2 \rightarrow A$
   7. $L - 1 \rightarrow L$

3. Determine the 8008 machine operation defined by the following hexadecimal codes (XX denotes NOP):

   a. XX
      62
      F3
      00
      XX

   b. XX
      C6
      XX

   c. XX
      28
      XX

   d. XX
      44
      BB
      00
      XX

   e. XX
      27
      XX

   f. XX
      13
      XX

   g. XX
      14
      08
      XX

   h. XX
      4B
      XX

4. Develop the flowcharts (logic and machine operation) to input 225 pieces of data from a sequential source via input register 3 and store the data in memory beginning at location $(1024)_{10}$. The advance signal for the sequential input is outputted at register (out) 18. Once entered, the order of sequence of the data is sequenced into descending order and stored in the same memory block, the highest number data in location $(1024)_{10}$ thru the lowest number data.

5. Develop the flowcharts (logic and machine operation) to perform the logical operation:

$$A \cdot (B + C \cdot D) \vee (D + C \cdot A) \cdot B = F$$

Input $A \rightarrow$ In $- 0$, $B \rightarrow$ In $- 1$, $C \rightarrow$ In $- 2$, and $D \rightarrow$ In $- 3$. Output $F \rightarrow$ Out $- 18$.

CHAPTER 13

Introduction to Programming Algorithms

The success or failure of the microprocessor-centered control system is a direct function of the programmer's use of programming solutions called *algorithms*. An algorithm may be defined as any software program necessary to perform a desired arithmetic or control function. Thus, the overall subject of the algorithm can be subdivided into a study of mathematical, logical, and process (or control) algorithms. The mathematical algorithm comprises a series of universal solution techniques that can be generally followed in the generation of virtually all software programs. The logic and process control algorithm, however, is unique to the system under consideration and cannot be generalized a priori. (*Note:* While absolute algorithms cannot be generated for a process-control program, specific techniques can be formulated for use in these programs.)

In developing the computer algorithm, it must be borne in mind that the microprocessor cannot perform any operation not listed in the operation code matrix. The mathematical algorithms which the microprocessor can perform must be developed from basically four fundamental operations, namely, ADD, SUB, RAL, and RAR. These four operations are combined with the process algorithms (primarily controlled looping) to create the complex mathematical functions of multiplication and division.

The technique of looping is of such major importance in virtually every complex algorithm that this process algorithm is developed before continuing further with either of the two major types of programming algorithm.

THE PROGRAMMED LOOP COUNTER

The programmed loop is the basic control program for all counting algorithms within the computer. If, for example, it is desired to input a series of 10 pieces of data into the computer, a single subroutine is written to perform the input operation and the subroutine repeated 10 times. The operational repeat may be accomplished by repeating the input subroutine ten times in ROM memory (although this is extremely wasteful of memory as well as being expensive) or by a simple routine incorporated to count the number of inputted data points and to exit out of the subroutine when 10 data points have been entered. This type of counting-control loop is programmed as follows.

1. The loop count (CNT) is programmed for the number of data points desired. In the illustration cited, the count is set to 10. The loop CNT is maintained in an on-chip register if possible.

2. The program to be included within the program loop is now written. For this illustration this program is left blank.

3. Following the program the loop-count register is decremented by 1 and the reduced value of CNT read into the accumulator. The new CNT value is compared to zero $(00)_{16}$. If the count (CNT) is greater than 0, a transfer is made back to step 2 and the program loop repeated. If the count equals 0, a transfer out of the program is made.

The flow diagram of steps 1–3 with its corresponding program is shown in Figure 13.1.

The placement of the decrement instruction is very critical in the loop algorithm. If this instruction is placed previous to the jump operation (ROM address 00, 19 of Figure 13.1), the computer only loads nine data points instead of 10 since it exits before completing the 10 loop. To correct this, the count register can be set to one count greater than the number to be counted, or the decrement instruction placed following the transfer instruction as indicated in the example. This latter technique is to be favored since the numerical value of the count can be entered into the program from an external source without concern over adjusting the count value.

The use of the decrement instruction instead of the increment instruction for counting is also convenient since the comparison of the count is always made with zero rather than a variable. However, while it is more convenient to control a loop by decrementing the counter, use of the algorithm for controlling the loop by incrementing the counter is not unreasonable. It does, however, utilize more program steps to perform the same algorithm and, therefore, is not pursued further here.

It should be apparent that the programming activity is one that does not have a *correct* or *wrong* way. The ultimate test for a program algorithm is its ability to perform the desired task within the allotted time period utilizing the minimum amount of ROM and RAM memory. If the algorithm meets these three criteria, it is to be regarded as a correct way. There may, however, be better ways of performing the same algorithm. To fail to do anything, however, while searching for the "best" algorithm is to miss the opportunity presented, for the "best" algorithm is illusive and always just around the corner.

To explore some of the possibilities of the loop-counting program, the following example can be used:

EXAMPLE.

Load 300 numbers into the microprocessor and locate the even numbers (2, 4, 6, · · ·) into memory beginning at location $100/_{10}$. In addition, all the odd input digits are added together in groups of two (i.e., $1 + 3$, $5 + 7$, $9 + 11$, etc.) and each sum outputted to register 10.

Solution

Figure 13.2 details the flow charts and machine-level program for this problem. Several important programming techniques are explored in this problem.

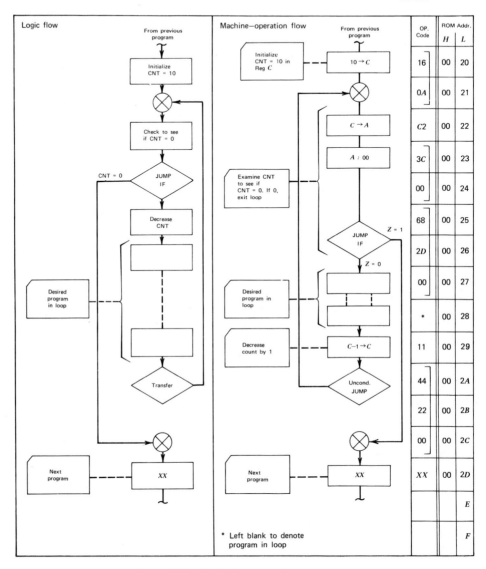

Figure 13.1. Basic loop-counting program.

First, the problem illustrates the necessity of thoroughly analyzing and evaluating the problem before attempting to solve it. The number of input digits (300) is a larger number than can be described by a single 8-bit byte (256). However, the program can be written in terms of a loop dealing simultaneously with four numbers. If this program technique is used, a total of 75 loops (rather than 300) is performed in order to input the desired 300 digits.

Second, the use of both incrementing (to determine the memory locations for each successive digit) and decrementing (loop counter) within the same program is illustrated.

Third, the technique for incrementing from one page of memory to the next page (increasing the higher-order address) is demonstrated.

The logic diagram indicates that the initial step in this solution is to set the loop counter at 75, the counter being located in register D. The decimal value of 75 is transformed to the hexadecimal code as follows:

$$75|_{10} = 2^6 + 2^3 + 2^1 + 2^0 = 0100, 1011 = 4,B|_{16}$$

The hexadecimal value of the memory location is determined by a combination of input-data count and initial memory location. Location $100|_{10}$ is located on page 00, so the H register (highest order address) is programmed to reflect the 00 page address. The location (L register) of address $100|_{10}$ is transformed as follows:

$$100|_{10} = 2^6 + 2^5 + 2^2 = 0110, 0100 = 6,4|_{16}$$

Once the D register has been initialized to CNT = 75 and the H and L registers initialized at location $100|_{10}$, the program loop can be initiated. The loop is entered by the JUMP command if CNT > 0. If CNT = 0 (register D = 0), the flag flip-flop Z is set to 1 and the loop program is exited and halts. If the flag flip-flop Z is set to 0 (CNT > 0), the main loop is initiated. The basic loop loads the first ADD number (data-digit 1) and places it in register B for temporary storage. The second (even) number is entered and placed into memory. The low-order memory address is incremented by one count and a JUMP instruction instituted if the Z-flag flip-flop is set to one. The only condition by which the flag flip-flop Z can be set to one at this point in the program is to reach the end of a memory page. Then the lower order address (L register) will be incremented by 1 from FF to 00. When this occurs and the flag flip-flop Z is set to 1, the higher-order address is incremented by 1. When the Z flip-flop remains 0, the main program is continued. The third number is entered (odd digit) and added immediately to the one stored in register B. The sum of these two numbers is outputted to output register 10. The fourth input number (even) is inputted and transferred to memory. Again the subroutine is entered to increment the lower address of memory and to check the Z flip-flop to determine if the memory end of page has been reached. If Z = 0, the main program is continued. The counter is decremented by 1 and the unconditional JUMP to loop is again performed.

THE ARITHMETIC ALGORITHMS

Addition and Subtraction

The algorithm for the mathematical functions of addition and subtraction are both contained within the Intel 8008 microprocessor chip. These basic functions of addition and subtraction can be programmed by a single op-code instruction located within the instruction-code matrix.

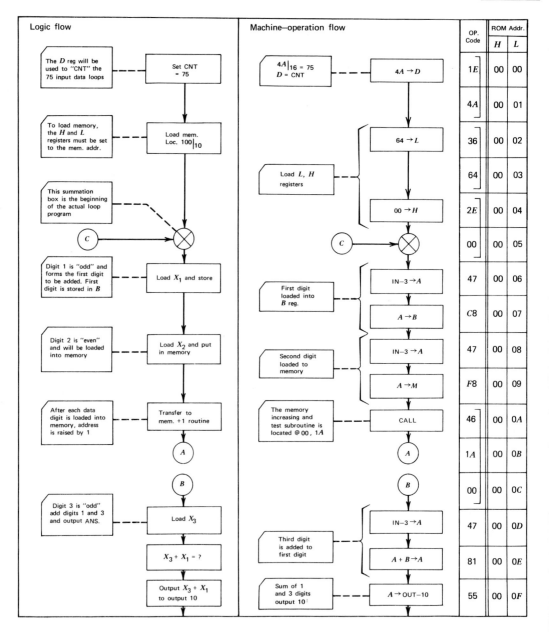

Figure 13.2. Loop-counting program.

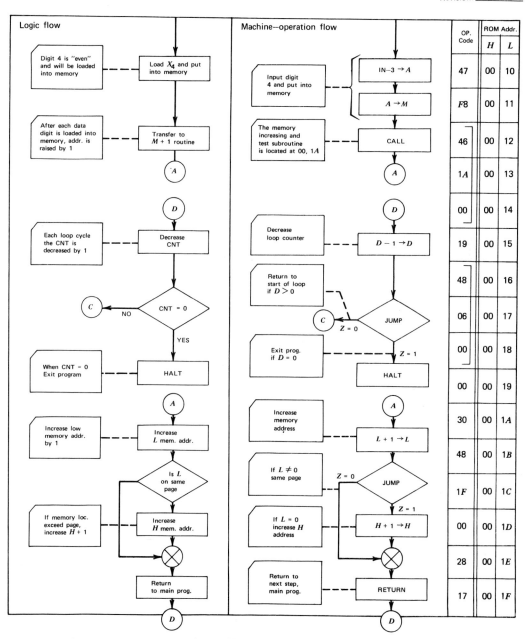

Figure 13.2. (continued)

Multiplication

The algorithm for multiplication is far more complex than a single op-code instruction. To understand the logical algorithm required to be performed by the computer, it is first necessary to understand the basis of multiplication in the decimal system.

Suppose the number 21 were to be multiplied by 42 without the use of mental multiplication. How would it be performed? The number 21 could be added to itself 42 times. Such an exercise would, however, be laborious and time-consuming. A second technique involves the knowledge of the decimal system of units, tens, hundreds, and higher powers that are used to create the decimal system. When the number 42 is examined, it is discovered that there are two "unit" digits and four "10" digits making up the number. In addition, the four "10" digits may be thought of as four "unit" digits times "10," which is equivalent to rotating the four digits to the left by one position. Thus, the number 42 could be viewed as:

> 2 unit digits
> and
> 4 unit digits rotated left one position

A number in the hundreds digit is equivalent to a unit digit rotated left by two positions ($X100$). This logic can be carried on indefinitely.

Since the computer can look at only one digit position at a time, assume that the "look" position is the unit digit of the multiplier. If the "10" digit is rotated right to locate it in the "units" position, the multiplicand must be rotated left (effectively multiplied by 10) to maintain mathematical equality.

In decimal form the multiplication of 42×21 would be:

1. Examine the unit digit position of multiplier.

$$4 \ 2$$
$$\uparrow \text{unit digit position}$$

2. Add the multiplicand with itself the number of times specified by the unit position digit (in this example, 2).

$$21$$
$$21$$
$$\overline{42} = \text{subtotal}$$

3. Rotate the multiplier right one position

$$— — → $$
$$4 \ \vdots \ 2$$
$$\uparrow$$
$$\text{unit digit position}$$

and rotate the multiplicand left one position.

$$← — —$$
$$2 \ 1 \ 0$$

4. Add the rotated multiplicand with itself the number of times specified by the number in the unit-digit position and sum this result with the previous subtotal.

$$210$$
$$210$$
$$210$$
$$210$$
$$\overline{840}$$
$$+42 \quad \text{(previous subtotal)}$$
$$\overline{882} = 21 \times 42$$

The multiplication of two binary digits by the computer is performed by the identical technique, except that binary numbers are used rather than decimal numbers. However, there are major limitations that must be observed when multiplying two numbers together in an 8-bit machine. The greatest of these limitations is the fact that the largest product that can be held within a single 8-bit word is 256. To be able to handle a larger product requires the programmer to program in double word lengths (16 bits) or greater. The programming complexities introduced when handling extended word lengths are extensive and should be carefully weighed before being undertaken. One technique that may be used to avoid the use of extended word lengths is the use of bit "weighting," which is discussed in Chapter 14 (The Candy-Factory Algorithm—An Example).

Assume that two binary numbers, m and n, are to be multiplied together; m will be located in B register while n is in C register, and n will be the multiplier. A counter will be placed in the E register to stop the multiplication after the 8-bits of the multiplier are examined. Register D will hold the running subtotal and the result of the multiplication will be outputted at register 12. The flow diagram and microprocessor program are shown in Figure 13.3.

Division

The algorithm for dividing one number by another is similar to that of multiplication except that the shifts are to the right instead of the left and the operation employed in the program is subtraction rather than addition. However, the resulting algorithm is much more complicated since there are significantly more comparisons to be made in order to determine the value of the answer. An analysis of the technique employed when two numbers are divided in the decimal system is required before the technique is translated into the binary system. Assume that the number 726 is to be divided by 3.

In the decimal system of division the number of 3s contained in the first digit (7) is determined along with the resultant remainder. This can be accomplished by subtracting 3 from 7 (leaving 4). The remainder is then compared with 3 and the process repeated if the remainder is greater than 3. The number of loops made before having a remainder less than 3 are counted yielding the first digit of the answer. The number (726) and the resulting answer (2) are both shifted one space to the left and the process again repeated. This process continues until the division is completed.

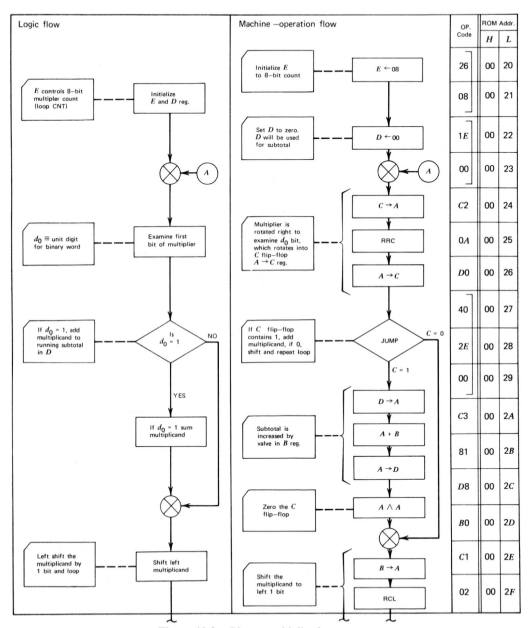

Figure 13.3. Binary multiplication program.

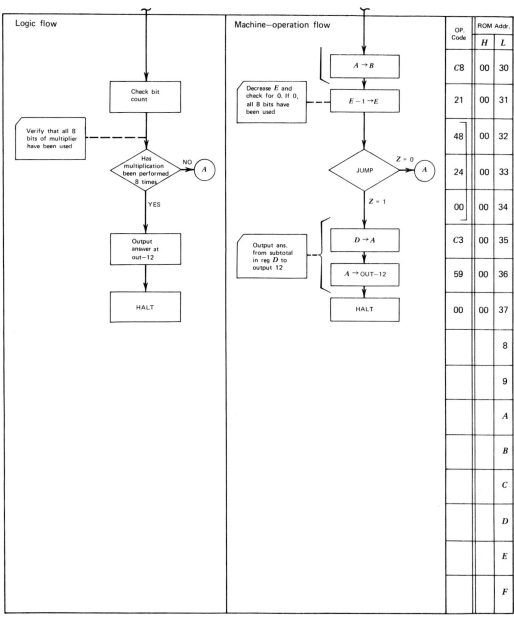

Logic flow	Machine—operation flow	OP. Code	ROM Addr. H	ROM Addr. L
	$A \rightarrow B$	$C8$	00	30
	Decrease E and check for 0. If 0, all 8 bits have been used — $E - 1 \rightarrow E$	21	00	31
Check bit count		48	00	32
Verify that all 8 bits of multiplier have been used — Has multiplication been performed 8 times — NO → A — JUMP $Z = 0$ → A	JUMP	24	00	33
YES	$Z = 1$	00	00	34
Output answer at out—12	Output ans. from subtotal in reg D to output 12 — $D \rightarrow A$	$C3$	00	35
	$A \rightarrow$ OUT—12	59	00	36
HALT	HALT	00	00	37
				8
				9
				A
				B
				C
				D
				E
				F

Figure 13.3. (continued)

159

Using the two numbers assumed (726 ÷ 3), the above process yields:

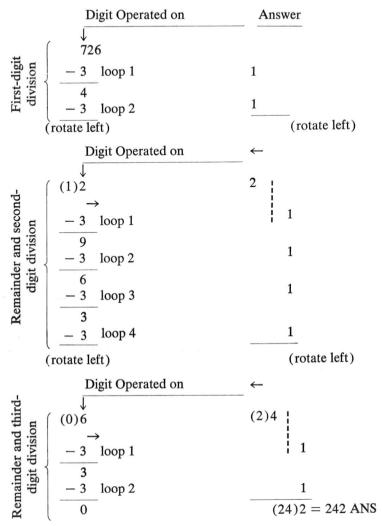

The above technique for the division of one decimal number by another must be modified slightly for binary division to account for the structure of the binary system. Since the binary system is a "1 − 0" structure, each "division" will be composed of a single subtraction loop that will result in a remainder greater than or equal to 0 (a valid subtraction) or less than 0 (an invalid subtraction). A valid subtraction yields a "1" binary digit value in the answer, whereas an invalid subtraction yields a "0" binary digit value. In addition, to keep track of the number of subtraction loops (shifts) performed, a counter must be included in the algorithm.

Translating the two decimal numbers into binary form yields:

$$726 = 1011010110, \quad 3 = 00000011$$

Assume the number 726 is placed into register C and 3 into register D. The total will be summed in register B. A magnitude counter will be maintained in register E. The logic steps to perform the binary division are:

1. Left shift the contents of register D until the D_7 bit value of 1 is obtained. Put this left justified number into register D. Count the number of left shifts required to reach a D_7 bit value of 1 and enter this value (n) plus 1 $(n + 1)$ into E.

2. Left shift the contents of register C until a D_7 bit value of 1 is obtained.

3. Subtract the left-shifted value of D found in step 1 from the left shifted value of C found in step 2. If answer is not negative, add 1 to register B. If answer is negative, add 0 to register B and replace A with value of C.

4. If $E > 0$, left shift register B one step, right shift register D one step. Decrement E register and test for zero.

5. Subtract value of D register (from step 4) from subtotal from step 3. If answer is negative, replace A with previous value of C register. If answer is not negative, add 1 to B register. Repeat step 4.

6. When value of the E register is equal to 0, output the value of register B as the answer.

The subtraction process followed by this algorithm is as follows:

		E (Counter)	B Register
Step 1:	$D = 0\ 0\ 0\ 0\ 0\ 0\ 0\ 1\ 1$	0	
	$\leftarrow$ left shift		
	$D = 1\ 1\ 0\ 0\ 0\ 0\ 0\ 0\ 0$	$8 + 1 = 9$	
	$8\ 7\ 6\ 5\ 4\ 3\ 2\ 1$		
Step 2:	$C = 1\ 0\ 1\ 1\ 0\ 1\ 0\ 1\ 1\ 0$	9	
	$\leftarrow$ left shift		
Step 3:	$C = 1\ 0\ 1\ 1\ 0\ 1\ 0\ 1\ 1\ 0$	8	$x\ x\ x\ x\ x\ x\ 0$
	$-D = 1\ 1\ 0\ 0\ 0\ 0\ 0\ 0\ 0\ 0$		$\leftarrow$ shift
	Answer <0		
Step 4:	$C = 1\ 0\ 1\ 1\ 0\ 1\ 0\ 1\ 1\ 0$	7	$x\ x\ x\ x\ x\ 0\ 1$
	$-D = x\ 1\ 1\ 0\ 0\ 0\ 0\ 0\ 0\ 0$		$\leftarrow$
Remainder $(R) =$	$1\ 0\ 1\ 0\ 1\ 0\ 1\ 1\ 0$		
Repeat 5:	$R = 0\ 1\ 0\ 1\ 0\ 1\ 0\ 1\ 1\ 0$	6	$x\ x\ x\ x\ 0\ 1\ 1$
	$-D = x\ x\ 1\ 1\ 0\ 0\ 0\ 0\ 0\ 0$		$\leftarrow$
	$R = 1\ 0\ 0\ 1\ 0\ 1\ 1\ 0$		
Repeat 6:	$R = 0\ 0\ 1\ 0\ 0\ 1\ 0\ 1\ 1\ 0$	5	$x\ x\ x\ 0\ 1\ 1\ 1$
	$-D = x\ x\ x\ 1\ 1\ 0\ 0\ 0\ 0\ 0$		$\leftarrow$
	$R = 1\ 1\ 0\ 1\ 1\ 0$		

Repeat 7: R = 0 0 0 0 1 1 0 1 1 0   4   x x 0̲ 1 1 1 1
          $-D$ = x x x x 1 1 0 0 0 0
          R =   0 0 0 1 1 0

Repeat 8: R = 0 0 0 0 0 0 0 1 1 0   3   x 0̲ 1 1 1 1 0
          $-D$ = x x x x x 1 1 0 0 0
          $R < 0$

Repeat 9: R = 0 0 0 0 0 0 0 1 1 0   2   x̲ 0 1 1 1 1 0 0
          $-D$ = x x x x x x 1 1 0 0
          $R < 0$

Repeat 10: R = 0 0 0 0 0 0 0 1 1 0   1   0̲ 1 1 1 1 0 0 1
          $-D$ = x x x x x x x 1 1 0
          R = 0 0 0 0 0 0 0 0 0 0

Repeat 11: R = 0 0 0 0 0 0 0 0 0 0   0   1̲ 1 1 1 0 0 1 0
          $-D$ = 0 0 0 0 0 0 0 0 1 1
          $R < 0$

Stop: answer = 242

The flowchart and program for this binary division algorithm are shown in Figure 13.4.

THE PROCESS-CONTROL ALGORITHM

The process-control algorithm differs from the arithmetic algorithm in that it controls the timing or operation of the computer/equipment under-control, whereas the arithmetic algorithm performs a complex mathematical operation. In the field of the dedicated microprocessor, more emphasis is focused on the process control algorithm than the various arithmetic algorithms.

The Indefinite Pause

A program illustration is developed to illustrate the technique of the computer "pausing," causing the computer to cease operation and wait for an external or internal command instruction to be provided. In reality, the microprocessor is never stopped, but is constantly performing the program dictated by the program memory. To pause or *tread* time until an event occurs before continuing the program requires a special microprogram. To pause for a *finite* time period before continuing the computer program requires a modification of this microprogram.

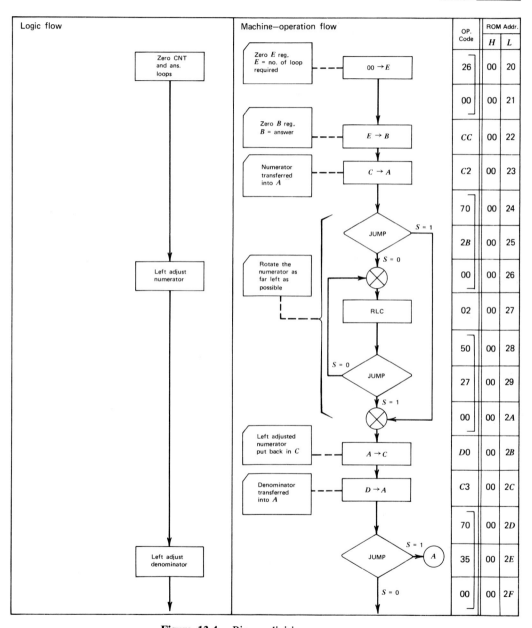

Figure 13.4. Binary division program.

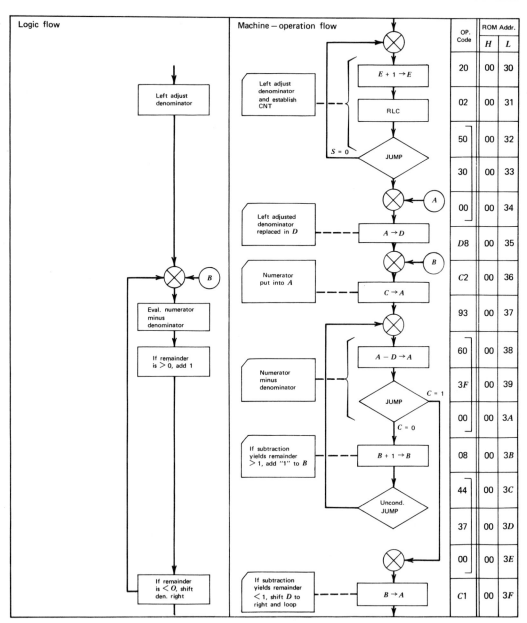

Figure 13.4. (continued)

164

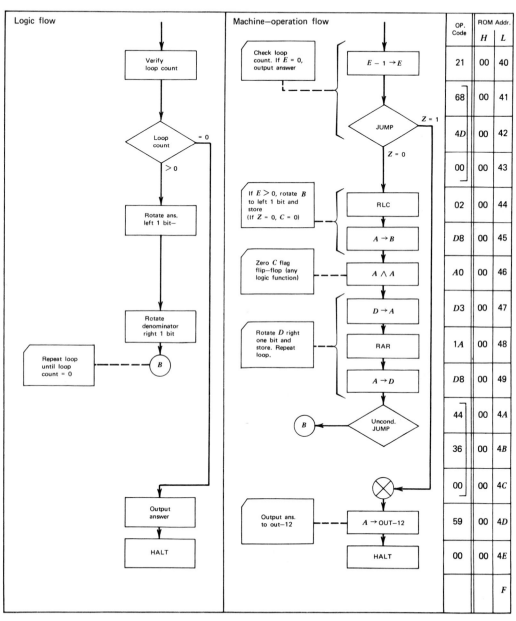

Figure 13.4. (continued)

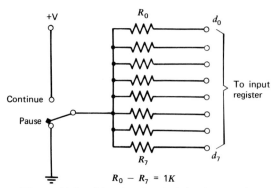

Figure 13.5. "Continue" switch implementation.

Initially assume that a computer pause algorithm is desired that can be exited (pause ceasing and the program continuing) by means of pushing a button (called CONTINUE) on the front panel. The continue button inserts the code *FF* into input 5 of the microprocessor. The *FF* code is extremely insensitive to noise since all 8 binary bits are required to be simultaneously set high. This input code is implemented as shown in Figure 13.5.

Once the processor has completed the first section of the desired program, the pause algorithm can be implemented by a series of no-operation codes referred to as NOP codes. These codes are signified by large dots on the hexadecimal instruction-code matrix sheets. The duration of these NOP instructions is determined by the programmer. In general, a short loop is desirable as it permits the fastest computer response to the CONTINUE switch command.

The main step in the pause loop is to input the data from input register 5 (either a 00 or *FF*) and then to compare it to the value *FF*. If the input value is unequal to *FF*, the program again looks at the input register and compares it with the desired data (*FF*). The microprocessor remains in this short loop until an *FF*-data byte is inputted by the command button. To the outside world, the computer has "paused." Figure 13.6 illustrates this type of program.

The minimum number of states (machine steps) required to complete one loop of *this* program is equal to 37. Each state takes from 1.25 μsec to 2 μsec to complete, yielding a minimum time delay of 46.25 μsec at 1.25 μsec/clock time duration (74 μsec at 2 μsec/clock time duration). The number of states required to perform a given instruction is indicated on the program-instruction explanation sheet.

The Fixed Time Delay

If a fixed time period delay is desired rather than a variable delay, a series of 12.5 to 20 μsec delays may be introduced into the program by the introduction of NOP instructions such as "load *A* into *A*" (instruction *C0*). This technique could prove laborious if a long time delay is desired. For long time delays a second technique is used that forms the delay time by means of a "pause" program controlled by a loop counter. The program for a 2 msec-delay is shown in Figure 13.7. Each loop

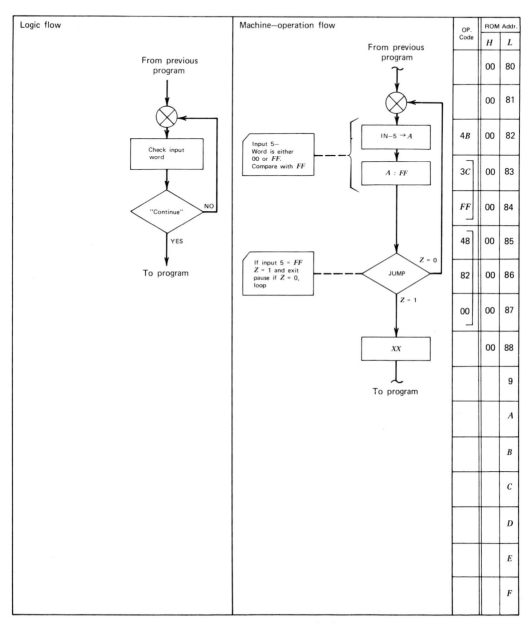

Figure 13.6. Pause loop—exit by CONTINUE signal program.

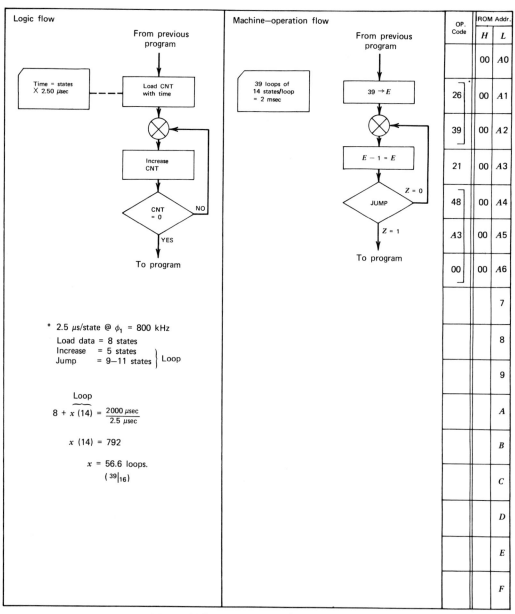

Figure 13.7. The 2-msec time-pause program.

of the program requires 14 states plus the original 8 states to enter and initialize the program. If an 800-kHz clock is used to control the microprocessor, each state requires 2.5 μsec to complete. The loop containing 14 states consumes a total time of 35 μsec. If 57 loops are made, a total elapsed time of 2.015 msec, including the loop initialization, is expended. The error in the 2-msec time-delay program illustrated will be approximately 15 μsec or 0.75%.

PROBLEMS

1. Input 239 data values into register 3 and store these values in memory beginning at memory location 400_{10}. Develop the logic and operation flowcharts and Intel 8008 machine-level program to perform this task.

2. The 239 numbers inputted into memory via problem 1 above are to be sequenced in ascending order. The value of the data ranges from 00_{16} to FF_{16}. The sequenced numbers are stored in a data block beginning at location 1024_{10}. Develop the logic and machine-operation flowcharts and the 8008 machine-level program to perform this task.

3. Develop the 8008 program for the "cruise computer" developed in problem 4 of Chapter 10. The speed in m.p.h. will be ≤ 120 m.p.h., the fuel-level input, ≤ 25 gallons, and the consumption rate, ≤ 20 gallons per hr.

4. A microprocessor is used to evaluate the musical performance of as many as 150 performers during a contest. Five separate musical parameters are in the judgment. They are, in order of priority: (a) accuracy, (b) difficulty, (c) tone, (d) technique, and (e) presentation. Each will be valued from 0 to 9, 9 being a perfect score. Develop a scoring computer for use by two or three judges, judging the same contestant simultaneously. The winner will be the contestant with the highest score, tie scores being broken by evaluation of the contestant having the highest overall priority scores taken from the highest priority to the lowest. Develop the 8008 machine program necessary to perform this task.

5. Develop the 8008 machine-level program required to find the average of n numbers. No prior information is known about "n" except that

$$\sum_0^n x < 10^9$$

where x is the value of the individual number.

CHAPTER 14

The Complete System Controllers
Hardware versus Software Programming

There exists a fine distinction between the portion of the microprocessor program to be programmed by means of external hardware and the portion of the program to be implemented by means of program software. The distinction is, of course, primarily a function of dollar cost and desired hardware speed. The distinction between hardware and software programming was once well defined and biased far in favor of the software program. However, with the emergence of the PLA, MSI analog chips, and sophisticated LSI technology chips, the bias has slowly been pushed toward the increased economics of using hardware programming supplemented by the software program.

The programmer is ultimately responsible for the evaluation of the trade-offs associated with the implementation techniques to be used in the development of the microprocessor system. For this reason he must be familiar with the basic hardware component devices available, the capability of these devices, and the costs associated with incorporating the external hardware into the processor design in order to adequately compare the hardware-programming option with the equivalent software-programming scheme.

THE TRAFFIC CONTROLLER—AN EXAMPLE

An example of this type of trade-off should suffice to illustrate the need for evaluating hardware versus software costs. Assume that a traffic controller is required for response to several external sensors. The roadway controlled is the complex intersection shown in Figure 14.1. The following conditions are to be met by the traffic controller:

1. Each lane will have traffic flow for 2* min.

2. Left-turn lane light is on for the initial 30* sec of the 2-min period.

3. During school transition periods (8:00-9:00 A.M., 3:00-4:00 P.M.), the A–A traffic lights will be on for 5* min, the others remaining at a 2*-min interval to permit school traffic flow and aid the children in crossing the streets.

* These times are approximate.

170

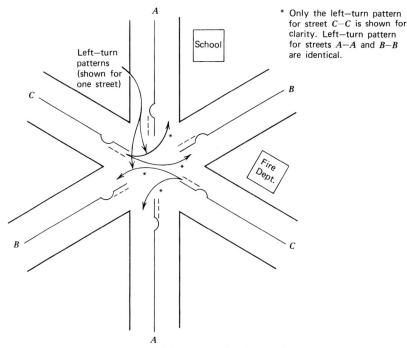

Figure 14.1. Controlled intersection.

4. During a fire call, all traffic lights will be red for 2* min to enable fire trucks to clear.

5. Walk requests will initiate directly after the 30*-sec "left turn" period and last until 30* sec before the traffic light is changed to red.

The Clock Function

The heart of the traffic controller resides in the timing function and sensor control. Two options remain open with respect to the timing function. They are

1. To program an internal clock based on the microprocessor's internal two-phase clock.

2. To use a commercially available external MSI chip clock such as the MM5312.

The required accuracy of the clock is established by the need to alter the traffic flow program during school transition periods. Since the traffic controller is expected to operate without human adjustment indefinitely, the clock accuracy must be virtually independent of component aging or inadvertant misadjustment. The accuracy of the fundamental frequency used to generate the two-phase clock for the microprocessor is relatively noncritical with respect to absolute frequency. However, if this frequency is used as a time standard, a crystal-controlled fre-

* These times are approximate.

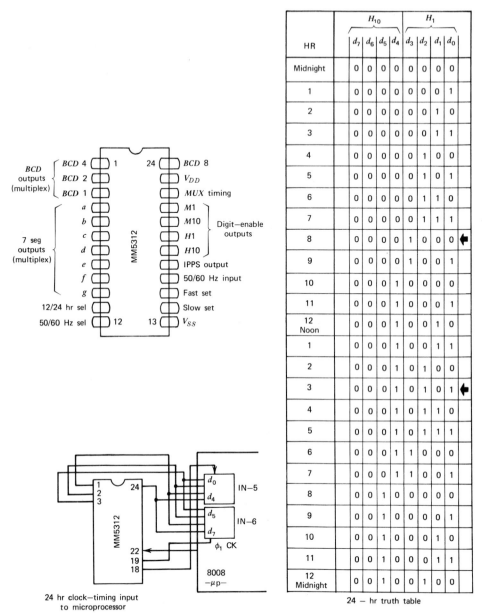

HR	H_{10}				H_1			
	d_7	d_6	d_5	d_4	d_3	d_2	d_1	d_0
Midnight	0	0	0	0	0	0	0	0
1	0	0	0	0	0	0	0	1
2	0	0	0	0	0	0	1	0
3	0	0	0	0	0	0	1	1
4	0	0	0	0	0	1	0	0
5	0	0	0	0	0	1	0	1
6	0	0	0	0	0	1	1	0
7	0	0	0	0	0	1	1	1
8	0	0	0	0	1	0	0	0
9	0	0	0	0	1	0	0	1
10	0	0	0	1	0	0	0	0
11	0	0	0	1	0	0	0	1
12 Noon	0	0	0	1	0	0	1	0
1	0	0	0	1	0	0	1	1
2	0	0	0	1	0	1	0	0
3	0	0	0	1	0	1	0	1
4	0	0	0	1	0	1	1	0
5	0	0	0	1	0	1	1	1
6	0	0	0	1	1	0	0	0
7	0	0	0	1	1	0	0	1
8	0	0	1	0	0	0	0	0
9	0	0	1	0	0	0	0	1
10	0	0	1	0	0	0	1	0
11	0	0	1	0	0	0	1	1
12 Midnight	0	0	1	0	0	1	0	0

24 − hr truth table

24 hr clock—timing input to microprocessor

Figure 14.2. The 24-hr clock timer.

quency would be required as a minimum. The cost for an accurate crystal-controlled frequency standard with temperature control (in areas where temperature extremes are expected) could be significant in respect to total hardware cost required for the clock algorithm.

An alternative to using an internal program-generated clock is the use of a commercial clock chip, programmed as a 24-hr clock. The output of a 24-hr clock can be directly utilized as the input timing reference to the microprocessor. Figure 14.2 illustrates the truth table of a 24-hr clock and the interconnection of the clock outputs to the computer input register. The cost of this chip is approximately $8.00 in small quantities and performs the same functional task as the internal program-generated clock requiring the addition of a crystal-controlled frequency standard and additional ROM memory to house the clock program. In both cases the use of a temperature-control element would be dictated if large temperature extremes were expected.

It is obvious that the use of the external clock is cost effective over the incorporation of a crystal controlled two-phase clock reference and the additional ROM program capacity to hold the counting algorithm. (The size of the additional ROM memory required to hold this algorithm is indeterminate since the physical counting algorithm cannot be programmed until all of the remaining traffic-control program has been generated because the traffic-control program execution time must be included in the clock algorithm.)

The Sensor Functions

The request-control sensors (walk, left turn, and fire) may be combined into signal input codes by the use of external hardware or individually decoded internally in the Intel 8008 chip. If the sensors are decoded by external hardware (by the use of a PLA or equivalent type of MSI circuitry) a minimum of a single-input register could be used to process a total of 256 external sensors. If the sensor is decoded internally within the 8008 chip, each sensor will require a dedicated input register. Figure 14.3 illustrates the general locations and type of sensor necessary to control this complex intersection.

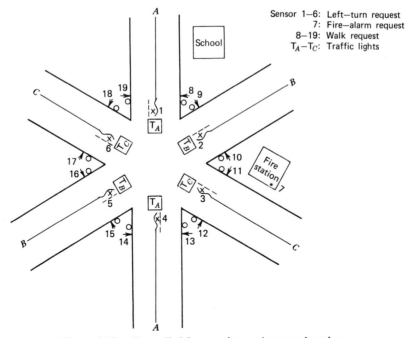

Figure 14.3. Controlled intersection and sensor locations.

The 19 required sensors exceed the eight available input registers and dictate the use of some form of external hardware implementation to sense the state of each sensor. The combined 19-sensor input requests could be formatted by use of a PLA to provide this status information to a single input register. It may, however, be advantageous to use more input registers to input the sensor data in order to minimize software programming.

Sensor Requests

The sensor requests will be as follows:

1. Walk: this request will be initiated by a pedestrian desiring to cross the street. This request is initiated by any of the four sensors controlling each street.
2. Left turn: the left-turn sensor is magnetically activated by the movement of a vehicle over an imbedded sensor in the roadway. Activation of either left-turn sensor will activate this request.
3. Fire call: this sensor is activated by the fire station.
4. School hours: a clock chip will perform the basic time-keeping chore. An internal software program will determine the time periods to enable the "school hours" subroutine.

Hardware Implementation of the Traffic Controller

The hardware implementation of the traffic controller must be established before the software program is written. This sequence is followed to enable the programmer to accurately determine the input and output codes required to perform the control process.

Output Control

Four output-control registers are required for the traffic-control application. Three of the output registers (out—20,out—21,out—22) are used to control the traffic lights on the three streets (A-A,B-B,C-C), while the fourth output register (out—23) is used to reset the various request flip-flops and the 2-min counter.

1. Traffic-light control: Figure 14.4 illustrates the traffic-light control hardware and truth table. The codes and hardware for each traffic light will be identical. The high-voltage SCR lamp circuits are not shown in Figure 14.4 since this circuitry will be included in the traffic light rather than in the controller. During the "walk" time period, the traffic light will also be green (for vehicular traffic). During this time a combined code (1100) will be outputted. During the left-turn time period, the traffic light will also be red for through vehicular traffic, and a combined code (1001) will be outputted.
2. Rest circuitry: eight functions (left turn A-A, left turn B-B, left turn C-C; walk A-A, walk B-B, walk C-C; fire alarm and timer) are required to have a program-controlled reset issued to them. The truth table for the reset register (out—23) is shown in Figure 14.5.

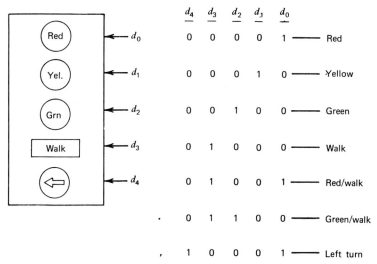

Figure 14.4. Traffic-light codes for output register.

Input-sensor Registers

While the full 8 bits of the input registers could be used to load the status of the total input requests, the ease of internal programming would lead the programmer to dedicate an input terminal to each multiple street request, the walk request occupying the d_7 input bit while the left turn request occupies the d_0 input bit. In this way the interrogation of the requests can be performed by sampling the sign bit (for walk request) or rotating right one bit and sampling the borrow (C) flip-flop (for left-turn request). Register In-1 will be used to input fire-alarm status; registers In-2, In-3, and In-4 will be used to input status of requests on streets A-A, B-B, and C-C respectively, registers In-5 and In-6 will input the 24-hr clock, and register In-7 will input the timer data.

The input PLA and register interface to the microprocessor controller is shown in Figure 14.6. Figure 14.7 details the PLA circuit configuration used for decoding the traffic controller.

	Output of Out—23							
	d_7	d_6	d_5	d_4	d_3	d_2	d_1	d_0
Reset fire alarm	0	0	0	0	0	0	0	1
Reset L.T. A-A	0	0	0	0	0	0	1	0
Reset L.T. B-B	0	0	0	0	0	1	0	0
Reset L.T. C-C	0	0	0	0	1	0	0	0
Reset walk A-A	0	0	0	1	0	0	0	0
Reset walk B-B	0	0	1	0	0	0	0	0
Reset walk C-C	0	1	0	0	0	0	0	0
Reset Timer	1	0	0	0	0	0	0	0

Figure 14.5. Truth table—reset circuitry.

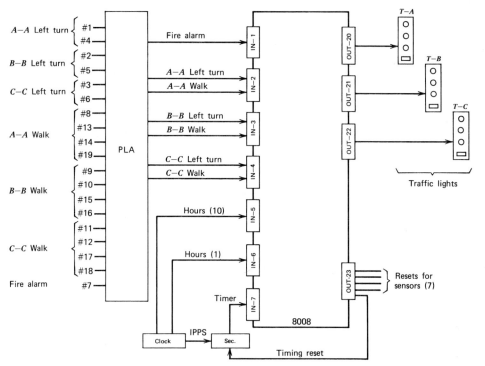

Figure 14.6. Traffic controller interface hardware.

The Traffic-control Logic Flowchart

The logic flowchart is virtually identical for each of the three streets, the total program being the serial combination of each of the three street subroutines. At the conclusion of the third subroutine, an unconditional return to the beginning will be made and the entire program repeated.

The logic program for the subroutine used to program the traffic flow on street *AA* is shown in Figure 14.8. The complete logic program for the controller repeats the essence of this program three times, once for each of the three streets. The complete machine language program for this traffic controller is shown in Figure 14.9.

The logic flowchart (Figure 14.9) is constructed as three separate programs, one for each main street. Several subroutines are provided for use in all three programs, including the subroutine for sensing the fire-alarm signal and the clock timer sequence. The basic logic flow sequence is as follows:

1. Set the 2-min timer.

2. Check sensor input for left-turn request. If a request is indicated, turn on left-turn light and hold for 30 sec. At the end of 30-sec period, turn off the left-

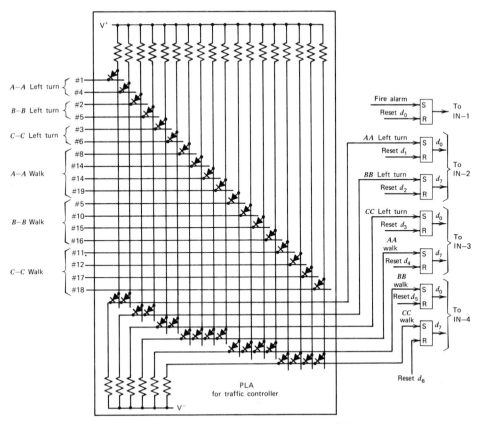

Figure 14.7. PLA Configuration and reset technique.

turn light and return to the main program. If no left-turn request has been received, continue with the main program.

3. Turn on green light for through traffic flow.

4. Check for walk requests for each of the two nonactive streets. If a request is sensed, issue the appropriate walk light signals and begin timing for 1 min (4 min for street *A-A* during school transition periods). At the end of this time period turn off the walk lights and return to the main program. If no walk requests are sensed, return to the main program.

5. Continue to check the 2-min timer. When it runs out, turn light red and continue main program.

6. Check the 24-hr clock to see if school transition periods are in progress: If so, extend walk periods on street *A-A*.

7. During all timing periods check for a fire alarm. This alarm overrides all other functions.

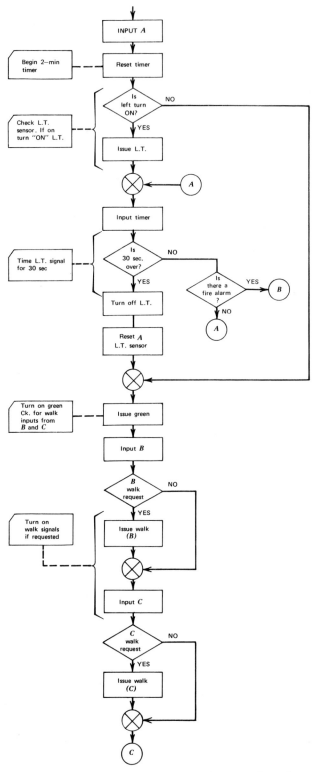

Figure 14.8. Street AA traffic-light controller logic flow diagram.

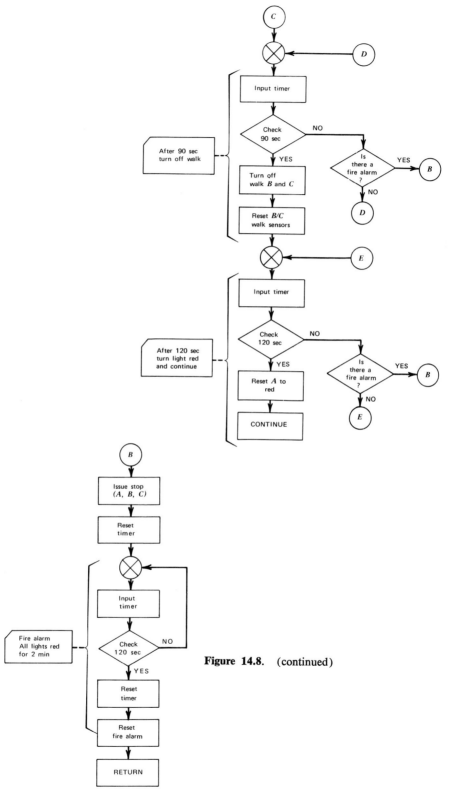

Figure 14.8. (continued)

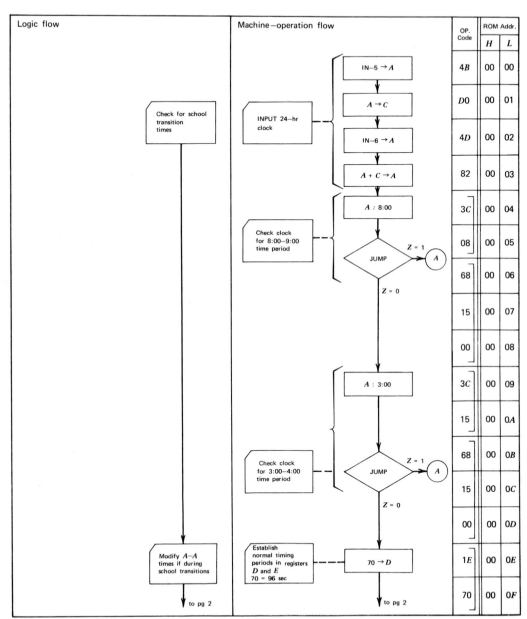

Figure 14.9. Traffic-light controller program.

180

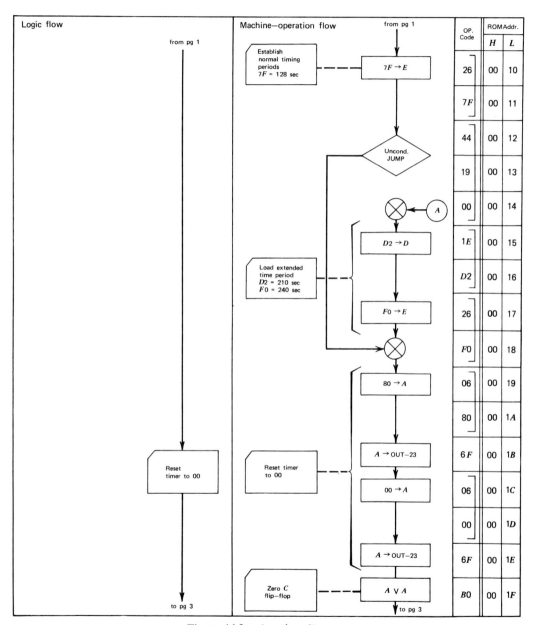

Figure 14.9.   (continued)

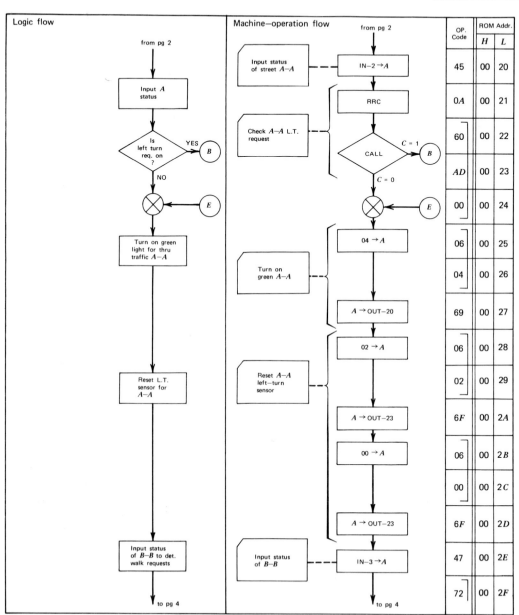

Figure 14.9. (continued)

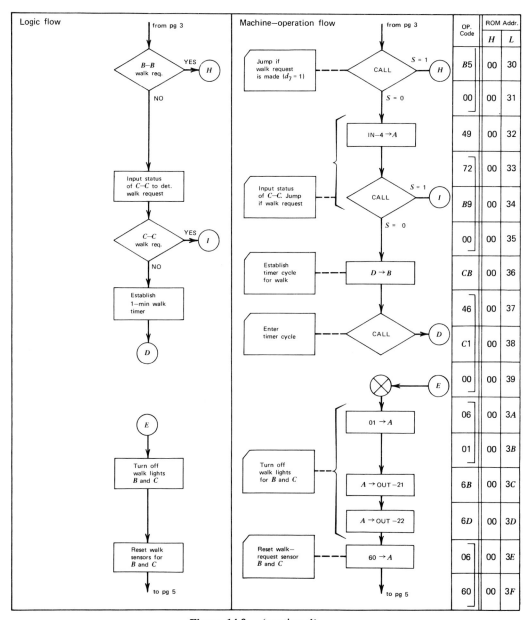

Figure 14.9. (continued)

Title: _Traffic Light Controller_   Page: _5_ of _15_

Programmer: _W. F. Leahy_   Date: _/ /_

                        Revision: _____

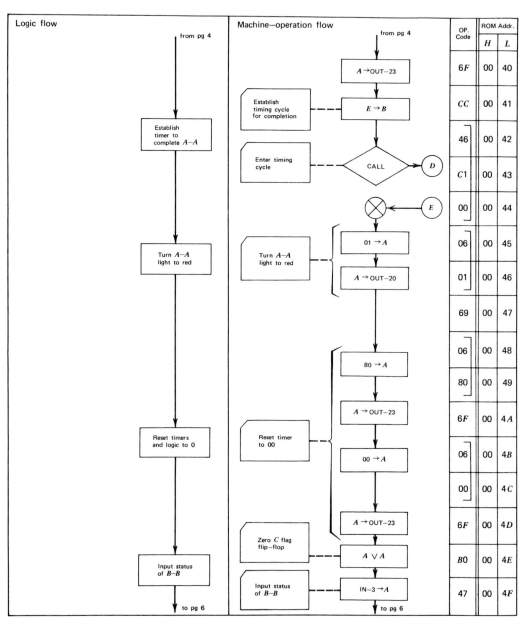

Figure 14.9. (continued)

184

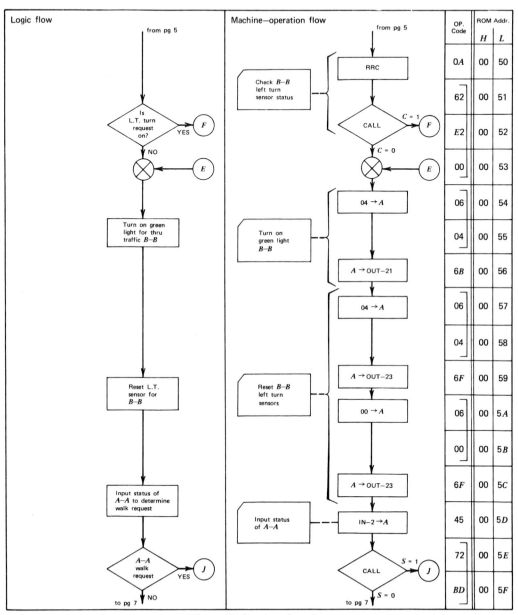

Figure 14.9. (continued)

185

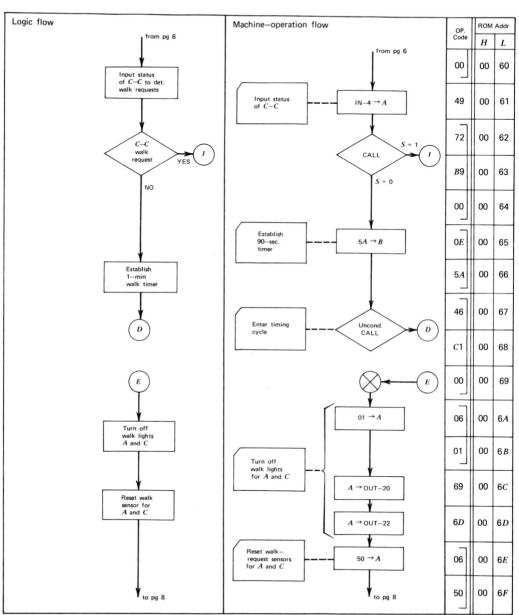

Figure 14.9. (continued)

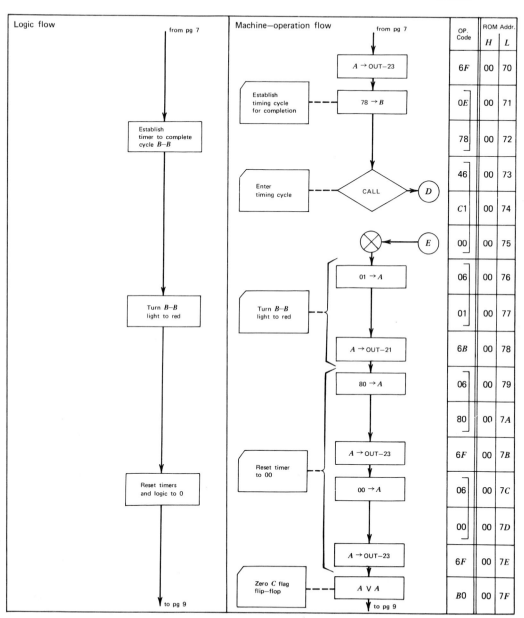

Figure 14.9. (continued)

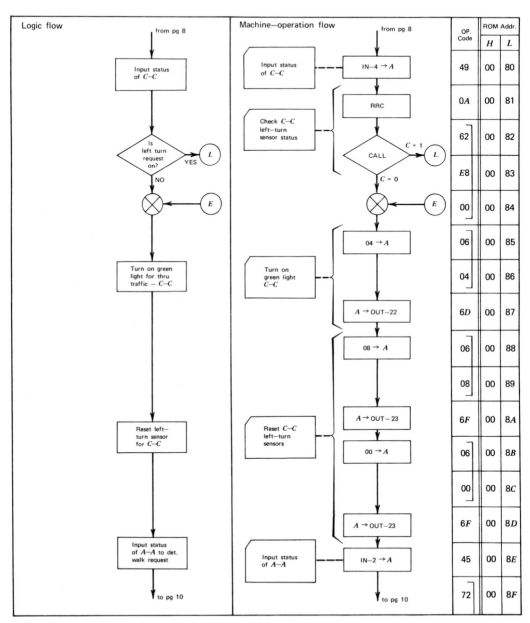

Figure 14.9. (continued)

188

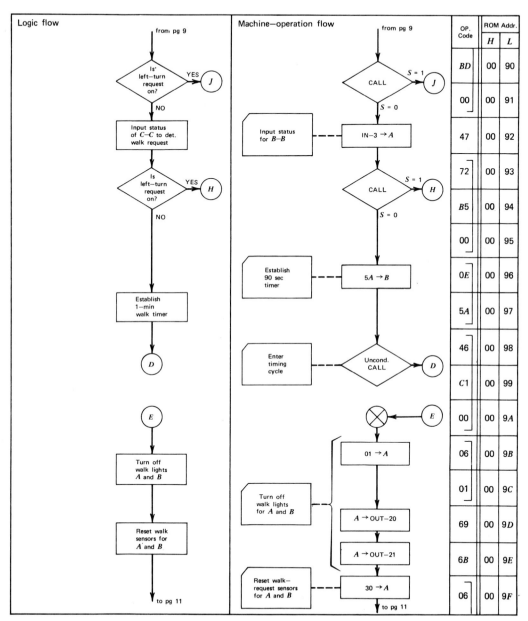

Figure 14.9. (continued)

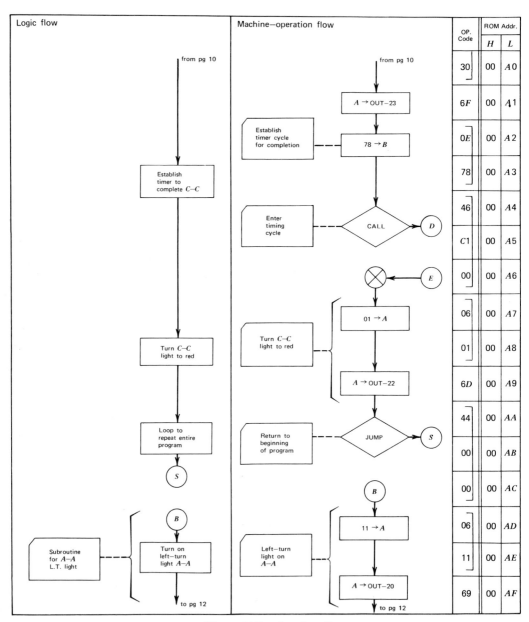

Figure 14.9. (continued)

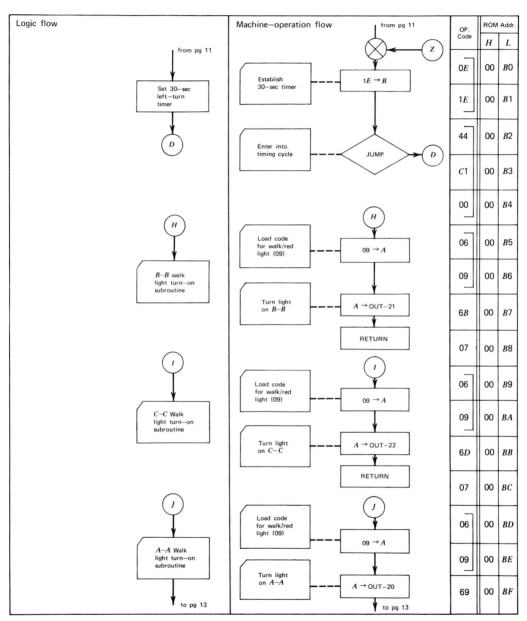

Figure 14.9. (continued)

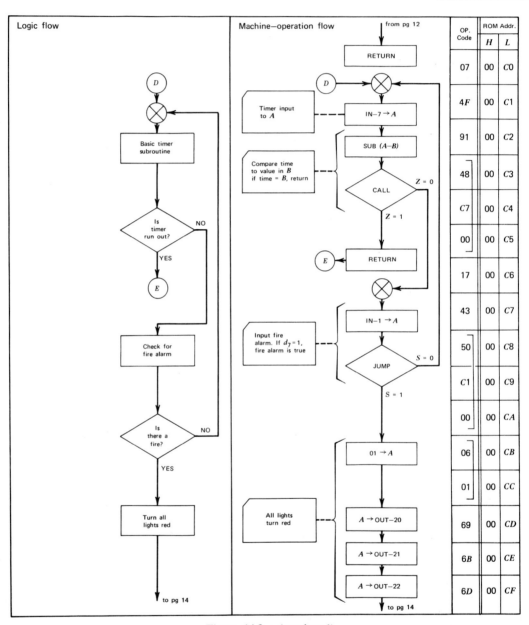

Figure 14.9. (continued)

192

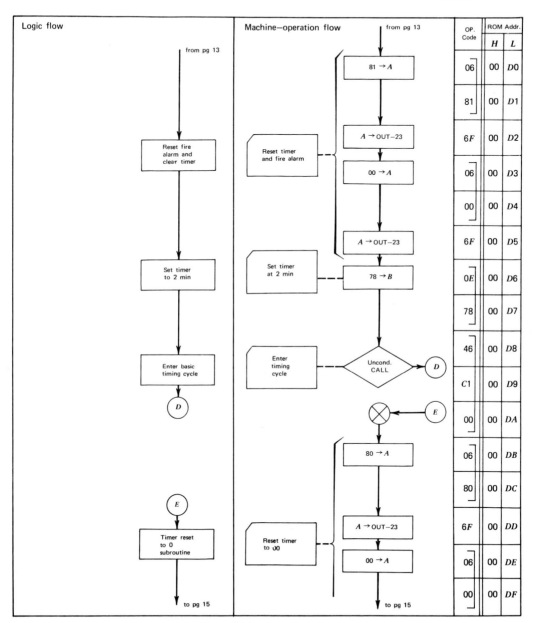

Figure 14.9. (continued)

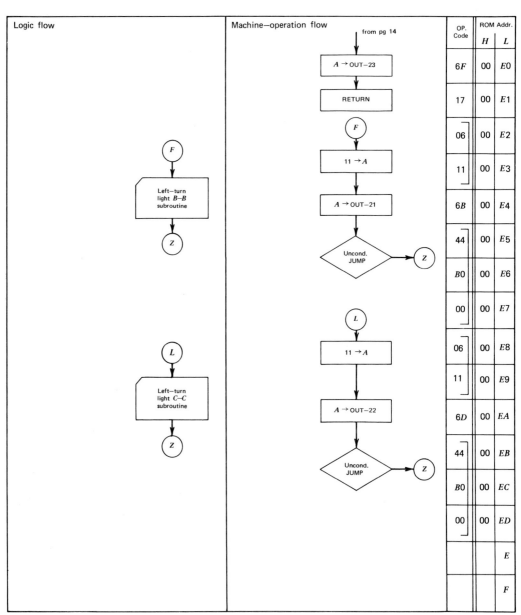

Figure 14.9. (continued)

194

THE CANDY-FACTORY ALGORITHM—AN EXAMPLE

The primary distinction between the arithmetic, logic, and the process-control algorithms is that the latter is dedicated to the task of controlling hardware performance, whereas the other two perform theoretical, mathematical, or logical manipulations. When the programmer enters into the realm of process-control programming, he is liberated from many of the programming restraints levied on the nonprocess-control programmers. For example, the programmer developing an arithmetic algorithm dealing with weight in ounces and pounds is required to define his units consistent with those definitions previously established by other programmers in the field; in other words, he must define weight in ounces and pounds. The process-control programmer, on the other hand, is required merely to define his units in a manner consistent with the hardware task before him, such as a 4-oz unit. Such a technique is referred to as programming in *weighted* units or *scaling*.

As an example of a process-control microprocessor system and the flexibility permitted in the process-control algorithm, consider the following design problem. The customer of a bulk candy producer has ordered a large quantity of 1-lb boxes of candy fudge. The contract is written, however, that boxes weighing less than this are unacceptable and must be repacked. Boxes weighing more than 1 lb will be accepted, but the supplier will lose on the cost of the overweight candy as the buyer will only pay for 1 lb. The candy manufacturer has agreed that he will pack his candy boxes to weigh 16 oz ($+0.1$ oz, -0.0 oz).

Figure 14.10 illustrates the candy-packaging station to be controlled by the microprocessor. Eight boxes of candy are simultaneously packaged at one station. Figure 14.11 illustrates the overall logic flow diagram for the microprocessor. Each piece of candy to be packaged enters the packaging station and is weighed, the candy having been precut to approximately 1-oz squares. However, because of variations in the density and height of the candy, the squares can vary in weight.

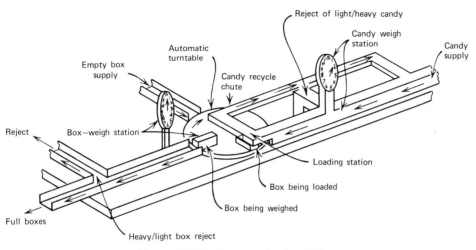

Figure 14.10. Candy-packaging station.

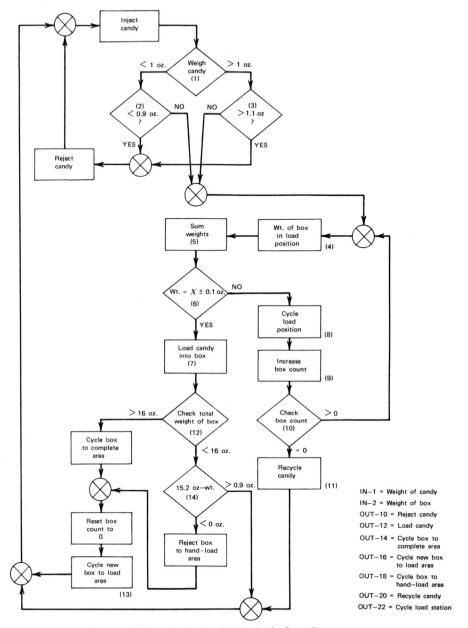

Figure 14.11. Candy-box loader logic flow diagram.

IN—1 = Weight of candy
IN—2 = Weight of box
OUT—10 = Reject candy
OUT—12 = Load candy
OUT—14 = Cycle box to complete area
OUT—16 = Cycle new box to load area
OUT—18 = Cycle box to hand—load area
OUT—20 = Recycle candy
OUT—22 = Cycle load station

If a piece of candy varies more than ± 0.1 oz, it will be rejected (logic decisions 2 and 3) and packaged by hand at a different station. If the weight of the candy is within 1 ± 0.1 oz, the candy will be packaged at this station. The weight of the partially filled box of candy under the loading position is measured (step 4) and added to the weight of the piece of candy to be loaded (step 5), the total weight being compared to determine if the new weight of the box with the candy added is within the desired ± 0.1-oz limit (step 6). If the total weight is within ± 0.1 oz, the candy is loaded into the box (step 7). If, however, the box is already overweight by >0.1 oz, a "no" response to the weight summation (step 6) would be recorded and the box rotated from under the loading position and a new box substituted (step 8). A running count is now initiated to determine how many boxes have been weighed with the same piece of candy being loaded (step 9). This running count is initialized at 00 when a piece of candy is loaded, and is incremented each time a box cannot be loaded. Step 10 subtracts the incremented box count from eight (the number of boxes being loaded) and, if all eight boxes are found to weigh incorrectly with a piece of candy, the candy is recycled to the end of the station line for reweighing at a later time (step 11). A new piece of candy is weighed (step 1) and the process cycle reinitiated.

Once the candy is loaded into a box (step 7), the box is weighed again (step 12) and its weight compared to the desired 16-oz limit. If it weighs ≥ 16 oz, the box is moved to the final packing area and an empty box substituted in its place (step 13). A new piece of candy is weighed (step 1) and the process is reinitiated. If the candy box weighs less than 16 oz, a final check is made to verify if the weight is less than 15.1 oz (thereby being able to accept one more piece of candy without exceeding the weight limit (step 14). If the weight is less than 15.1 oz, the box is maintained at its station to be recycled. If it is found to weigh more (a technical impossibility assuming that the other weighings were performed correctly), the box will be rejected to the special handling area to have the final piece of candy selected from the "light-weight" pieces already rejected in step 2. Figure 14.12 details the 8008 program for the mechanization of the above program.

Of primary importance in this program mechanization is the bit value *weighting* assigned to this process control algorithm. Throughout the description of the program emphasis has been placed on the ± 0.1-oz weight limitation that is required to be maintatined. Since the only interface to the microprocessor is the machine interface (weighing machines), the LSB unit value may be defined as equal to 0.1 oz. As a result of this weighted input value, the programmer has no need to program in decimal-value numbers. A piece of candy is measured as 9–11 units rather than 0.9 to 1.1 oz. The full box of candy now weighs 160^{+1}_{-0} units rather than $16^{+0.1}_{-0}$ oz. In short, the process-control programmer can often eliminate many of the tacky programming problems associated with unit dimensions, which universally confront the arithmetic and logic algorithm programmer.

Title: <u>*Candy factory*</u>

Programmer: <u>*W. F. Leahy*</u>

Date: / /

Revision:

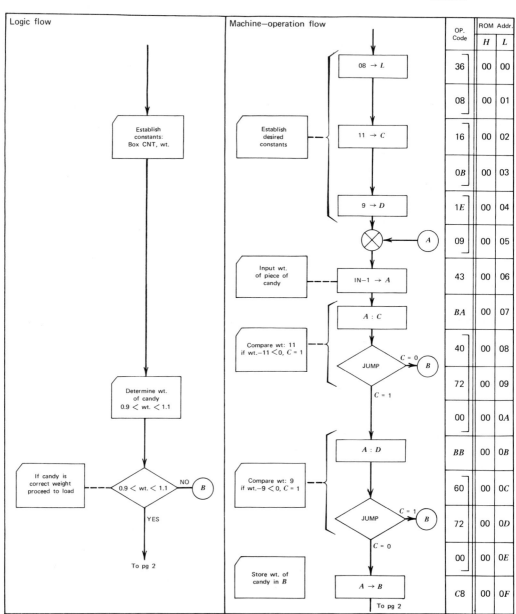

Figure 14.12. Candy-factory program.

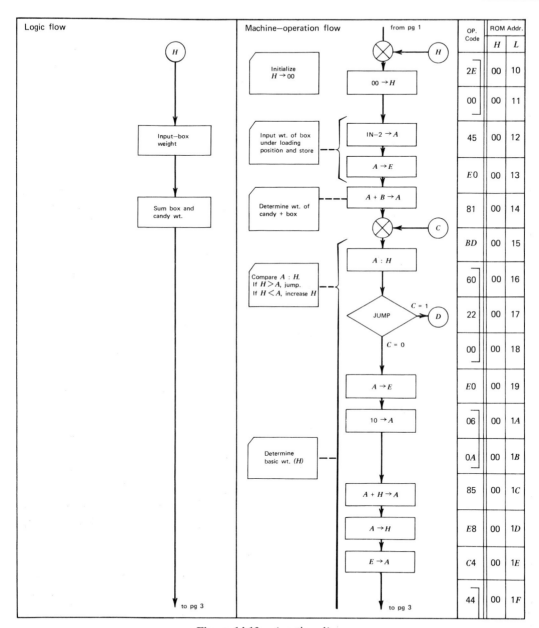

Figure 14.12. (continued)

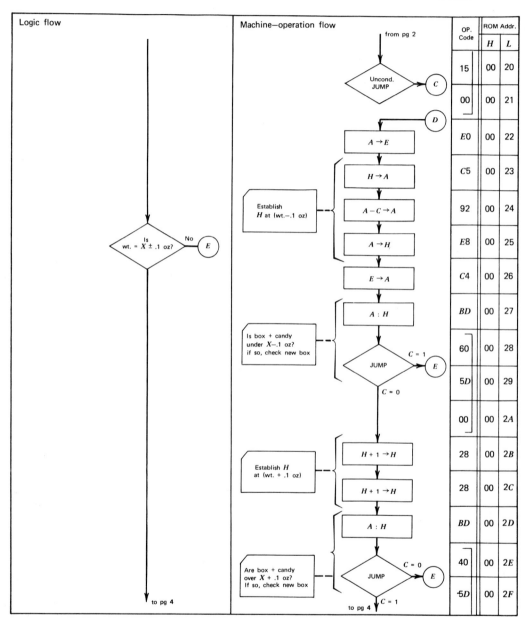

Figure 14.12. (continued)

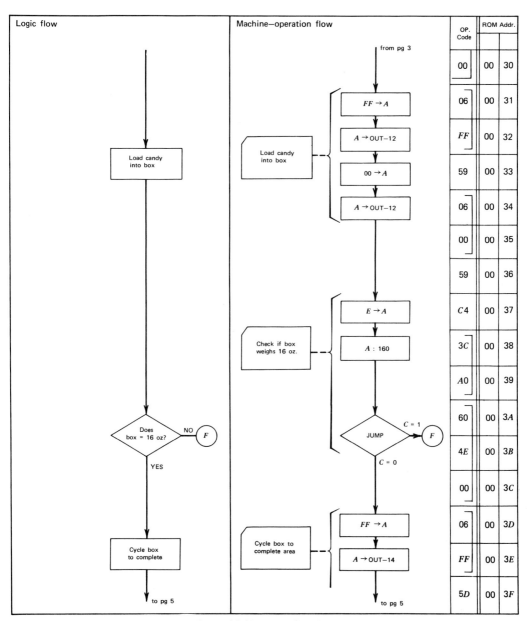

Figure 14.12. (continued)

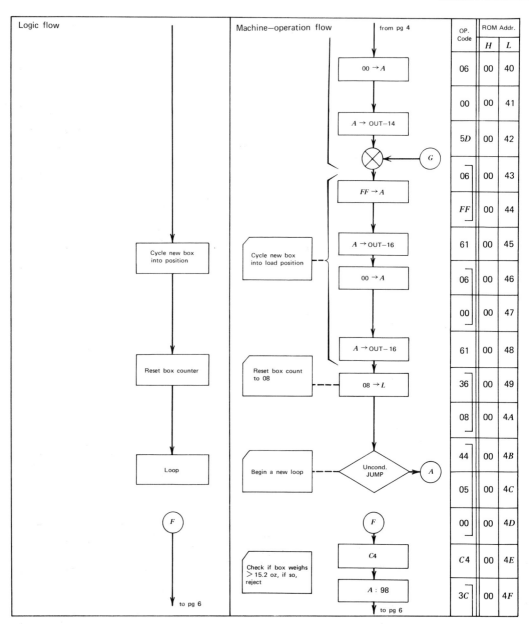

Figure 14.12. (continued)

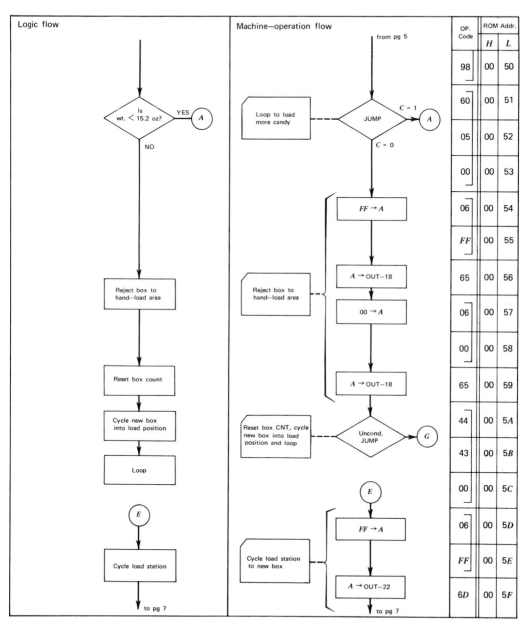

Figure 14.12. (continued)

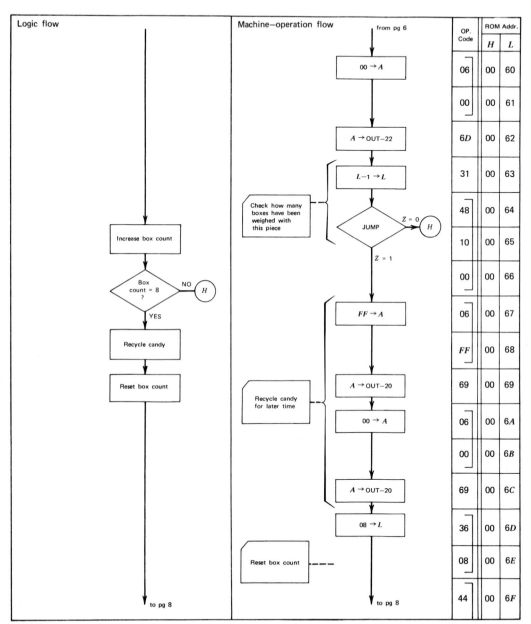

Figure 14.12. (continued)

204

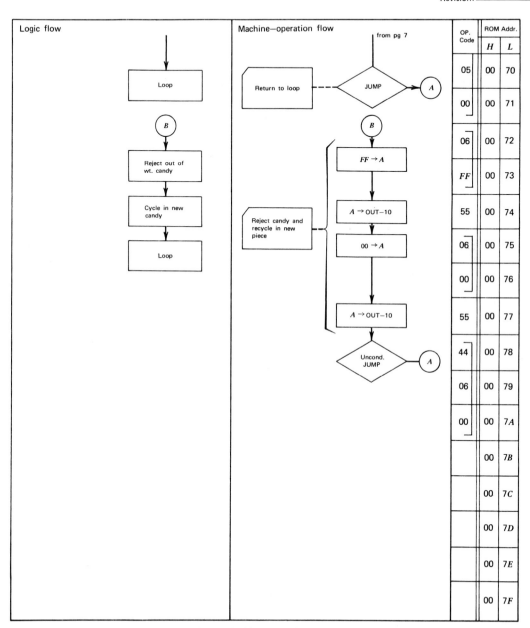

		OP. Code	ROM Addr.	
			H	L

Logic flow

Loop

B

Reject out of wt. candy

Cycle in new candy

Loop

Machine—operation flow

from pg 7

Return to loop — JUMP — A

B

FF → A

A → OUT—10

Reject candy and recycle in new piece

00 → A

A → OUT—10

Uncond. JUMP — A

OP. Code	H	L
05	00	70
00	00	71
06	00	72
FF	00	73
55	00	74
06	00	75
00	00	76
55	00	77
44	00	78
06	00	79
00	00	7A
	00	7B
	00	7C
	00	7D
	00	7E
	00	7F

Figure 14.12. (continued)

205

PROBLEMS

1. Program fully the problem:

$$E = a! \quad \text{where} \quad 0 < a < 5$$

   (Do not merely multiply!)

2. At a convention, a door prize will be given away at uniform intervals. These intervals correspond to the sum of 250 computer units as determined for the individuals entering the convention. A computer unit is determined by the running unit count as follows: Age of the person entering minus the years that person has worked at his present job, or his age divided by two, whichever is less. When the running count reaches *exactly* 250 computer units, the individual entering wins the door prize. If the running count *exceeds* 250, the prize will be awarded to the next individual entering the convention. The winner will be notified by means of a flashing red light. Develop the flowcharts and 8008 program to perform this task. The person's age will be inputted at In 1 and the years worked at the present job at In 2. The output signifying a winner will be at Out 15.

3. Develop an 8008 program to provide change from 1¢ to 99¢ with the fewest number of coins possible. The available coins will be inputted as follows: 1¢ → In 1, 5¢ → In 2, 10¢ → In 3, 25¢ → In 4, and 50¢ → In 5. The corresponding change will be outputted as follows: 1¢ → Out 12, 5¢ → Out 13, 10¢ → Out 14, 25¢ → Out 15, and 50¢ → Out 16. The value of the change required will be inputted at In 6.

4. Two trains travel on separate tracks until they reach Middletown. The trains share a common track from Middletown to Hartford some 30 mi apart. The tracks redivide at Hartford. The trains can be traveling at any speed between 30 m.p.h. and 60 m.p.h. depending on the cargo they are carrying. Under *no* circumstance is train A to stop. Develop the flowcharts and 8008 computer program to determine if train B will be required to stop.
   Sensors *A* and *B* are set to determine exact train velocity in m.p.h. Sensor *C* responses to the presence or absence of a train on the common track. If train B is stopped, it *must* be started again as soon as it is safe. Develop the 8008 flowcharts and machine program to perform this task.

5. The microprocessor is to be used to perform a "point-of-intercept" calculation between two spacecrafts. The input data are determined from an on-board radar that gives data in three-dimensional coordinates in m.p.h. relative to your own spacecraft. These three inputs will be: V_x → In 1, V_y → In 2, and V_z → In 3. The steering commands to your spacecraft will be outputted as follows: V_x → Out 20, V_y → Out 21, and V_z → Out 22. Determine the flowcharts and 8008 program to perform this calculation.

CHAPTER 15

Beyond The Intel 8008

Once a firm understanding of the basic components and programming techniques of the microprocessor are attained, the question may well be asked, "What next?" The answer lies primarily in the hands of the control engineer.

The future of the microprocessor lies in the diversity of its use and the ingenuity of the designer/programmer. While the microprocessor is presently enjoying its "day in the sun" in "smart" point-of-sale terminals, electronic equipment and in the myriad television games that have become popular beyond most expectations, the final test of the microprocessor will be found in its *staying* power, its achievement in performing mundane tasks as the unnoticed servant. In this role of the dedicated servant, the Intel 8008 microprocessor chip, with all of its imperfections and shortcomings, is a viable control-processor candidate, for it is both inexpensive and powerful. As the engineer begins to evaluate the every-day control tasks currently being marginally performed by use of out-dated control techniques, there opens a tremendous market, as yet untapped due to the lack of creative technology. Some of these markets may well include the following:

- Efforts are presently placed on the conservation of energy in the automobile by means of computer-controlled ignition–carburation systems. Microprocessor-controlled instrumentation is also needed to indicate maximum economy speed ranges, to define required maintenance intervals, and to provide information such as cruise range based on remaining gas supply and present consumption.

- Heating and cooling control systems for the home and office could bring about the constant-temperature room, supplying heat or refrigeration to the rooms requiring it (based on shifts in sunlight, etc.) and closing off vents to rooms that are still at the desired temperature.

- Control systems for automatic sun tracking platforms containing solar cells would permit direct sunlight for the maximum period of the day and would follow the sun's seasonal shifting position. These platforms would maximize the power attainable from the solar-cell technology.

- The use of the microprocessor in the leisure enjoyment fields is virtually without limit. The changeover of personal and commercial communication systems from the present AM/FM transmission techniques to digital transmission opens new horizons for increasing fidelity as well as decreasing complexity and cost.

- Personal communication systems offer the camper a new freedom to roam without fear of becoming lost. Speech-compression techniques formerly reserved for highly complex installations can be incorporated into personal communicators to permit time multiplexing of the airways to a degree not dreamed of before, enabling the use of the personal two-way radio for every person without the limitations so prevalent today.

- The personal communicator cited above will of course prove a boon to personal safety from injury or attack.
- Medical monitors can be designed to constantly monitor a heart patient and warn him of impending danger even while he is leading a normal life and performing his daily routine.
- A home or store security system can be introduced to not only sense an intrusion or fire but also to telephone the proper authorities. The security system can be programmed to automatically turn on and off lights, control watering the garden, and so on.

This list provides but a superficial glance at the potential future of the microprocessor control technology. The first three listed subparagraphs pertain to *energy conservation,* the next two pertain to *increased leisure enjoyment,* and the final three subparagraphs relate to *increased safety.* Not all of these control systems can be performed with a basic microprocessor such as the Intel 8008. Since its introduction in the early 1970s, many more advanced processor chips have been introduced that contain more inherent programming capability and speed than the first generation 8008 microprocessor. While the majority of these newer microprocessor systems are tailored to fill a unique marketing need, there are two second-generation general-purpose microprocessors that should be noted, namely, the Intel 8080 microprocessor chip and the Motorola M6800 microprocessor.

THE INTEL 8080 MICROPROCESSOR SYSTEM

The most noticeable physical difference between the 8008 and 8080 devices is the large size of the 40-pin package of the 8080 compared with the minute 18-pin package of the 8008 (see Figure 15.1). This increased pin capacity permits the 8080 to have independent 16-bit tristate address lines, 8-bit data bus lines and 18 decode/states lines. This is in contrast to the complex time multiplexing of the data bus lines for data and addressing required in the 8008. The 16-bit internal latch permits direct addressing of 65,536 bytes of memory by the 8080 without use of external address registers. The address lines also communicate with the input/output devices with an 8-bit address byte permitting the 8080 to use up to 256 input/output devices.

The 8080 Program-instruction Set

The 8080 microprocessor program-instruction set uses virtually the same instruction set as the 8008 **plus** a number of new instructions. However, the instruction code designations (locations) of these identical instructions have been changed in the 8080 code matrix.

Pin configuration

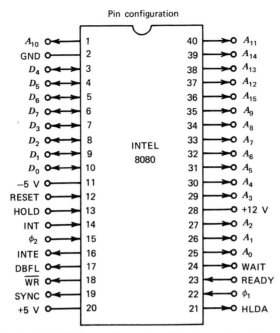

Figure 15.1. Intel 8080 pin configuration. (Courtesy of Intel Corporation.)

The 8080 instructions identical in operation to those of the 8008 are:

MOV	MVI $\langle B_2 \rangle$		
INR	DCR		
ADD	ADI $\langle B_2 \rangle$		
ADC	ACI $\langle B_2 \rangle$		
SUB	SUI $\langle B_2 \rangle$		
SBB	SBI $\langle B_2 \rangle$		
NDA	NDI $\langle B_2 \rangle$		
XRA	XRI $\langle B_2 \rangle$		
ORA	ORI $\langle B_2 \rangle$		
CMP	CPI $\langle B_2 \rangle$		

RLC   RRC   RAL   RAR

JMP $\langle B_2 \rangle$ $\langle B_3 \rangle$* JC $\langle B_2 \rangle$ $\langle B_3 \rangle$* JNC $\langle B_2 \rangle$ $\langle B_3 \rangle$* JZ $\langle B_2 \rangle$ $\langle B_3 \rangle$* JNZ $\langle B_2 \rangle$ $\langle B_3 \rangle$*

HLT   JP $\langle B_2 \rangle$ $\langle B_3 \rangle$* JM $\langle B_2 \rangle$ $\langle B_3 \rangle$* JPE $\langle B_2 \rangle$ $\langle B_3 \rangle$* JPO $\langle B_2 \rangle$ $\langle B_3 \rangle$*

All of these instruction details are the same as the 8008.

* D_6 & D_7 of $\langle B_3 \rangle$ are used as active high-level address bits.

Intel 8080 Instructions Differing from 8008 in Detail Only

Mnemonic	Bytes	Cycles	Description of Operation
CALL $\langle B_2 \rangle$ $\langle B_3 \rangle$	3	5	$[SP - 1] [SP - 2] \leftarrow (PC)$, $(SP) = (SP) - 2 (PC) \leftarrow \langle B_3 \rangle \langle B_2 \rangle$ Transfer the content of *PC* to the pushdown stack in memory addressed by the register *SP*.

Mnemonic	Bytes	Cycles	Description of Operation
			The content of SP is decreased by 2. Jump unconditionally to the instruction located in memory location addressed by byte 2 and byte 3 of the instruction.
CC $\langle B_2 \rangle$ $\langle B_3 \rangle$	3	3/5	If (carry) = 1 $[SP - 1]$ $[SP - 2] \leftarrow PC$, $(SP) = (SP) - 2$, $(PC) \leftarrow \langle B_3 \rangle \langle B_2 \rangle$; otherwise $(PC) = (PC) + 3$
CNC $\langle B_2 \rangle$ $\langle B_3 \rangle$	3	3/5	If (carry) = $\emptyset$ $[SP - 1]$ $[SP - 2] \leftarrow PC$, $(SP) = (SP) - 2$, $(PC) \leftarrow \langle B_3 \rangle \langle B_2 \rangle$; otherwise $(PC) = (PC) + 3$
CZ $\langle B_2 \rangle$ $\langle B_3 \rangle$	3	3/5	If (zero) = 1 $[SP - 1]$ $[SP - 2] \leftarrow PC$, $(SP) = (SP) - 2$, $(PC) \leftarrow \langle B_3 \rangle \langle B_2 \rangle$; otherwise $(PC) = (PC) + 3$
CNZ $\langle B_2 \rangle$ $\langle B_3 \rangle$	3	3/5	If (zero) = $\emptyset$ $[SP - 1]$ $[SP - 2] \leftarrow PC$, $(SP) = (SP) - 2$, $(PC) \leftarrow \langle B_3 \rangle \langle B_2 \rangle$; otherwise $(PC) = (PC) + 3$
CP $\langle B_2 \rangle$ $\langle B_3 \rangle$	3	3/5	If (sign) = $\emptyset$ $[SP - 1]$ $[SP - 2] \leftarrow PC$, $(SP) = (SP) - 2$, $(PC) \leftarrow \langle B_3 \rangle \langle B_2 \rangle$; otherwise $(PC) = (PC) + 3$
CM $\langle B_2 \rangle$ $\langle B_3 \rangle$	3	3/5	If (sign) = 1 $[SP - 1]$ $[SP - 2] \leftarrow PC$, $(SP) = (SP) - 2$, $(PC) \leftarrow \langle B_3 \rangle \langle B_2 \rangle$; otherwise $(PC) = (PC) + 3$
CPE $\langle B_2 \rangle$ $\langle B_3 \rangle$	3	3/5	If (parity) = 1 $[SP - 1]$ $[SP - 2] \leftarrow PC$, $(SP) = (SP) - 2$, $(PC) \leftarrow \langle B_3 \rangle \langle B_2 \rangle$; otherwise $(PC) = (PC) + 3$
CPO $\langle B_2 \rangle$ $\langle B_3 \rangle$	3	3/5	If (parity) = $\emptyset$ $[SP - 1]$ $[SP - 2] \leftarrow PC$, $(SP) = (SP) - 2$, $(PC) \leftarrow \langle B_3 \rangle \langle B_2 \rangle$; otherwise $(PC) = (PC) + 3$
RET	1	3	$(PC) \leftarrow [SP]$ $[SP + 1]$ $(SP) = (SP) + 2$. Return to the instruction in the memory location addressed by the last values shifted into the pushdown stack addressed by SP. The content of SP is incremented by 2.
RC	1	1/3	If (carry) = $\emptyset$ $(PC) \leftarrow [SP]$, $[SP + 1]$, $(SP) = (SP) + 2$; otherwise $(PC) = (PC) + 1$
RNC	1	1/3	If (carry) = 1 $(PC) \leftarrow [SP]$, $[SP + 1]$, $(SP) = (SP) + 2$; otherwise $(PC) = (PC) + 1$
RZ	1	1/3	If (zero) = 1 $(PC) \leftarrow [SP]$, $[SP + 1]$, $(SP) = (SP) + 2$; otherwise $(PC) = (PC) + 1$
RNZ	1	1/3	If (zero) = $\emptyset$ $(PC) \leftarrow [SP]$, $[SP + 1]$, $(SP) = (SP) + 2$; otherwise $(PC) = (PC) + 1$
RP	1	1/3	If (sign) = $\emptyset$ $(PC) \leftarrow [SP]$, $[SP + 1]$, $(SP) = (SP) + 2$; otherwise $(PC) = (PC) + 1$
RM	1	1/3	If (sign) = 1 $(PC) \leftarrow [SP]$, $[SP + 1]$, $(SP) = (SP) + 2$; otherwise $(PC) = (PC) + 1$

Mnemonic	Bytes	Cycles	Description of Operation
RPE	1	1/3	If (parity) = 1 $(PC) \leftarrow [SP], [SP + 1], (SP) = (SP) + 2$; otherwise $(PC) = (PC) + 1$
RPO	1	1/3	If (parity) = $\emptyset$ $(PC) \leftarrow [SP], [SP + 1], (SP) = (SP) + 2$; otherwise $(PC) = (PC) + 1$
RST	1	3	$[SP - 1] [SP - 2] \leftarrow (PC), (SP) = (SP) - 2$ $(PC) \leftarrow (00000000 \quad 00AAA000)$ This new RST instruction code will be "11AAA111".
IN $\langle B_2 \rangle$	2	3	$(A) \leftarrow$ (Input data) At T_1, time of third cycle, byte 2 of the instruction, which denotes the I/O device number, is sent to the I/O device through the address lines,† and the INP status information, instead of MEMR, is sent out. Therefore, the data multiplexor to the 8080 can be switched from memory to input. New data for the accumulator are loaded from the data bus when DBFL control signal is active. The condition flip-flops are not affected.
OUT $\langle B_2 \rangle$	2	3	(Output data) $\leftarrow (A)$ At T_1, time of the third cycle, byte 2 of the instruction, which denotes the I/O device number, is sent to the I/O device through the address lines,† and the OUT status information is sent out. The content of the accumulator is made available on the data bus when the $\overline{\text{WR}}$ control signal is $\emptyset$.

* D_6 and D_7 of $\langle B_3 \rangle$ are used.

† The device address appears on $A_7 - A_\phi$ and $A_{15} - A_8$.

The Entirely New 8080 Instructions*

Mnemonic	Bytes	Cycles	Description of Operation
LXI B, $\langle B_2 \rangle$ $\langle B_3 \rangle$	3	3	$(C) \leftarrow \langle B_2 \rangle; (B) \leftarrow \langle B_3 \rangle$ Load byte 2 of the instruction into C. Load byte 3 of the instruction into B.
LXI D, $\langle B_2 \rangle$ $\langle B_3 \rangle$	3	3	$(E) \leftarrow \langle B_2 \rangle, (D) \leftarrow \langle B_3 \rangle$ Load byte 2 of the instruction into E. Load byte 3 of the instruction into D.
LXI H, $\langle B_2 \rangle$ $\langle B_3 \rangle$	3	3	$(L) \leftarrow \langle B_2 \rangle, (H) \leftarrow \langle B_3 \rangle$ Load byte 2 of the instruction into L. Load byte 3 of the instruction into H.
LXI SP, $\langle B_2 \rangle$ $\langle B_3 \rangle$	3	3	$(SP)_L \leftarrow \langle B_2 \rangle, (SP)_H \leftarrow \langle B_3 \rangle$ Load byte 2 of the instruction into the lower-order 8-bit of the stack pointer and byte 3 into the higher-order 8-bit of the stack pointer.

* Copyright Intel Corporation—reprinted by permission.

Mnemonic	Bytes	Cycles	Description of Operation
PUSH A	1	3	$[SP-1] \leftarrow (A), [SP-2] \leftarrow (F), (SP) = (SP) - 2$ Save the contents of A and F (5-flags) into the pushdown stack addressed by the SP register. The contents of SP is decremented by 2. The flag word appears as follows: D_0: CY_2 (Carry) D_1: 1 D_2: Parity (even) D_3: $\emptyset$ D_4: CY_1 D_5: $\emptyset$ D_6: Zero D_7: MSB (sign)
PUSH B	1	3	$[SP-1] \leftarrow (B)\ [SP-2] \leftarrow (C), (SP) = (SP) - 2$
PUSH D	1	3	$[SP-1] \leftarrow (D)\ [SP-2] \leftarrow (E), (SP) = (SP) - 2$
PUSH H	1	3	$[SP-1] \leftarrow (H)\ [SP-2] \leftarrow (L), (SP) = (SP) - 2$
POP A	1	3	$(F) \leftarrow [SP], (A) \leftarrow [SP+1], (SP) = (SP) + 2$ Restore the last values in the pushdown stack address by SP into A and F. The content of SP is increased by two.
POP B	1	3	$(C) \leftarrow [SP], (B) \leftarrow [SP+1], (SP) = (SP) + 2$
POP D	1	3	$(E) \leftarrow [SP], (D) \leftarrow [SP+1], (SP) = (SP) + 2$
POP H	1	3	$[L] \leftarrow [SP], (H) \leftarrow [SP+1], (SP) = (SP) + 2$
STA $\langle B_2 \rangle$ $\langle B_3 \rangle$	3	4	$[\langle B_3 \rangle \langle B_2 \rangle] \leftarrow (A)$ Store the accumulator content into the memory location addressed by bytes 2 and 3 of the instruction.
LDA $\langle B_2 \rangle$ $\langle B_3 \rangle$	3	4	$(A) \leftarrow [\langle B_3 \rangle \langle B_2 \rangle]$ Load the accumulator with the content of the memory location addressed by bytes 2 and 3 of the instruction.
XCHG	1	1	$(H) \leftrightarrow (D)\ (E) \leftrightarrow (L)$ Exchange the contents of registers H and L and registers D and E.
XTHL	1	5	$(L) \leftrightarrow [SP], (H) \leftrightarrow [SP+1]$ Exchange the contents of registers H, L, and the last values in the pushdown stack addressed by registers SP. The SP register itself is not changed. $(SP) = (SP)$.
SPHL	1	1	$(SP) \leftarrow (H)\ (L)$ Transfer the contents of registers H and L into register SP.
PCHL	1	1	$(PC) \leftarrow (H)\ (L)$ JUMP INDIRECT
DAD SP	1	3	$(H)\ (L) \leftarrow (H)\ (L) + (SP)$ Add the content of register SP to the content of registers H and L and place the result into registers H and L. If the overflow is generated, the carry flip-flop is set; otherwise, the carry flip-flop is reset. The other condition flip-flops are not affected. This is useful for addressing data in the stack.

Mnemonic	Bytes	Cycles	Description of Operation
DAD B	1	3	$(H)(L) \leftarrow (H)(L) + (B)(C)$
DAD H	1	3	$(H)(L) \leftarrow (H)(L) + (H)(L)$ (double precision shift left H and L)
DAD D	1	3	$(H)(L) \leftarrow (H)(L) + (D)(E)$
STAX B	1	2	$[(B)(C)] \leftarrow (A)$ Store the accumulator content in the memory location addressed by the content of registers B and C.
STAX D	1	2	$[(D)(E)] \leftarrow (A)$ Store the accumulator content into the memory location addressed by the content of registers D and E.
LDAX B	1	2	$(A) \leftarrow [(B)(C)]$ Load the accumulator with the content of the memory location addressed by the content of registers B and C.
LDAX D	1	2	$(A) \leftarrow [(D)(E)]$ Load the accumulator with the content of memory location addressed by the content of registers D and E.
INX B	1	1	$(B)(C) \leftarrow (B)(C) + 1$ The content of register pair B and C is incremented by 1. All of the condition flip-flops are not affected.
INX H	1	1	$(H)(L) \leftarrow (H)(L) + 1$ The content of registers H and L is incremented by 1. All of the condition flip-flops are not affected.
INX D	1	1	$(D)(E) \leftarrow (D)(E) + 1$
INX SP	1	1	$(SP) \leftarrow (SP) + 1$
DCX B	1	1	$(B)(C) \leftarrow (B)(C) - 1$
DCX H	1	1	$(H)(L) \leftarrow (H)(L) - 1$
DCX D	1	1	$(D)(E) \leftarrow (D)(E) - 1$
DCX SP	1	1	$(SP) \leftarrow (SP) - 1$
CMA	1	1	$(A) \leftarrow \overline{(A)}$ The content of accumulator is complemented. The condition flip-flops are not affected.
STC	1	1	$(\text{carry}) \leftarrow 1$ Set the carry flip-flop to 1. The other condition flip-flops are not affected.
CMC	1	1	$(\text{carry}) \leftarrow \overline{(\text{carry})}$ The content of carry is complemented. The other condition flip-flops are not affected.
INR A	1	1	$(A) \leftarrow (A) + 1$ The content of accumulator is incremented by 1. All of the condition flip-flops except carry are affected by the result.
DCR A	1	1	$[A] \leftarrow [A] - 1$ The content of accumulator is decremented by 1. All of the condition flip-flops except carry are affected by result.

213

Mnemonic	Bytes	Cycles	Description of Operation
DAA	1	1	Decimal adjust accumulator. The 8-bit value in the accumulator containing the result from an arithmetic operation on decimal operands is adjusted to contain two valid *BCD* digits by adding a value according to the following rules:

$$
\begin{array}{cccc}
7 & 4 & 3 & 0
\end{array}
$$

X	.	Y

Accumulator

If $(Y > = 1\emptyset)$ or (carry from bit 3) then $Y = Y + 6$ with carry to X digit.

If $(X > = 1\emptyset)$ or (carry from bit 7) or $[(Y = 1\emptyset)$ and $(X = 9)]$ the $X = X + 6$ (which sets the carry flip-flop).

Two carry flip-flops are used for this instruction. CY_1 represents the carry from bit 3 (the 4th bit) and is accessible as a 5th flag. CY_2 is the carry from bit 7 and is the usual carry bit.

All condition flip-flops are affected by this instruction.

Mnemonic	Bytes	Cycles	Description of Operation
SHLD $\langle B_2 \rangle$ $\langle B_3 \rangle$	3	5	$[\langle B_3 \rangle \langle B_2 \rangle] \leftarrow (L), [\langle B_3 \rangle \langle B_2 \rangle + 1] \leftarrow (H)$ Store the contents of registers H and L into the memory location addressed by bytes 2 and 3 of the instructions.
LHLD $\langle B_2 \rangle$ $\langle B_3 \rangle$	3	5	$(L) \leftarrow [\langle B_3 \rangle \langle B_2 \rangle], (H) \leftarrow [\langle B_3 \rangle \langle B_2 \rangle + 1]$ Load the registers H and L with the contents of the memory location addressed by bytes 2 and 3 of the instruction.
EI	1	1	Interrupt system enable.
DI	1	1	Interrupt system disable. The 8080 has the interrupt enable flip-flop (INTE), the INTE flip-flop can be set or reset by using the above-mentioned instructions. The INT signal will be accepted if the INTE is set. When the INT signal is accepted by the CPU, the INTE will be reset immediately.
INR M	1	3	$[M] \leftarrow [M] + 1$ The content of memory designated by registers H and L is incremented by 1. All of the condition flip-flops except carry are affected by the result.
DCR M	1	3	$[M] \leftarrow [M] - 1.$

The Hexadecimal 8080 Operation-Code Matrix

Figure 15.2 illustrates the Intel 8080 Hexadecimal Operation Code Matrix, and Figure 15.3 details the general location of the instruction groups. Like the 8008 operation code matrix, the 8080 matrix may be broken into three horizontal blocks and two vertical blocks. Section I extends from code 0 to 3, section II con-

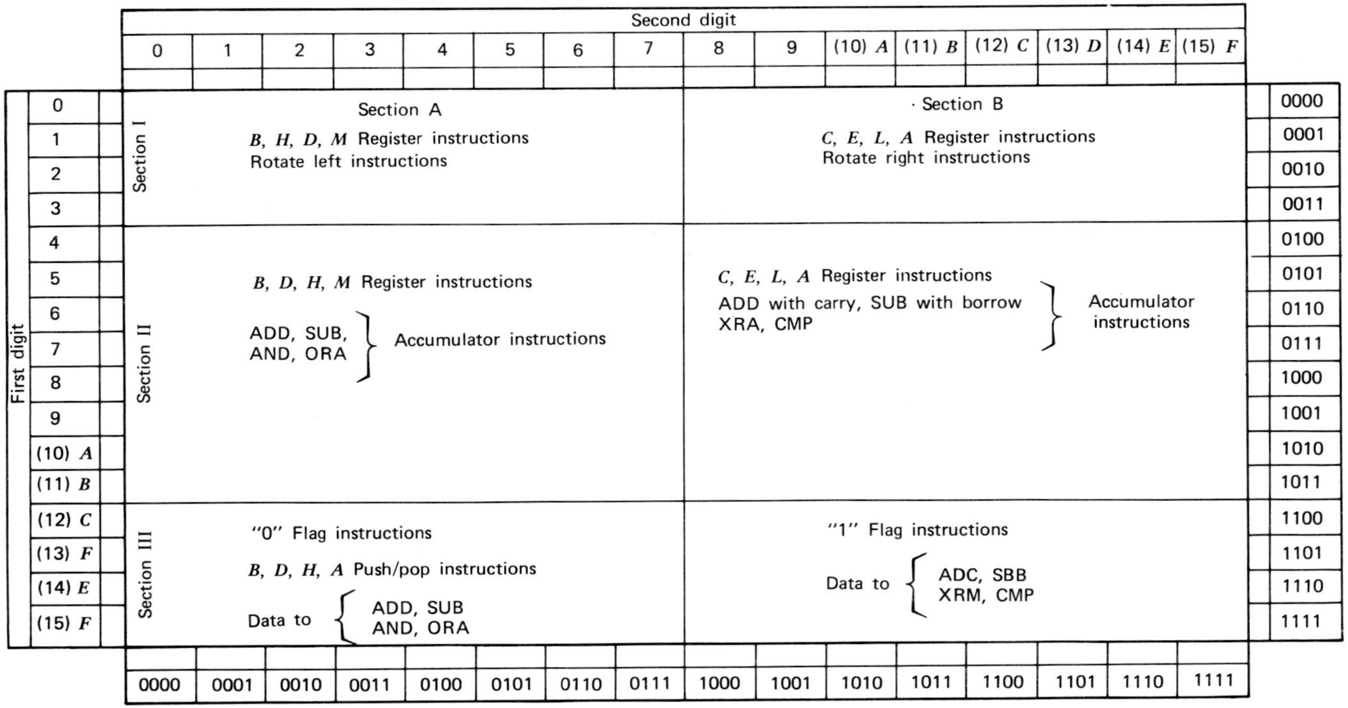

Second digit
0 1 2 3 4 5 6 7 8 9 (10) A (11) B (12) C (13) D (14) E (15) F
First digit
0 1 2 3 4 5 6 7 8 9 (10) A (11) B (12) C (13) F (14) E (15) F
Section I
Section A
B, H, D, M Register instructions
Rotate left instructions
Section B
C, E, L, A Register instructions
Rotate right instructions
Section II
B, D, H, M Register instructions
ADD, SUB,
AND, ORA
Accumulator instructions
C, E, L, A Register instructions
ADD with carry, SUB with borrow
XRA, CMP
Accumulator
instructions
Section III
"0" Flag instructions
B, D, H, A Push/pop instructions
Data to
ADD, SUB
AND, ORA
"1" Flag instructions
Data to
ADC, SBB
XRM, CMP
0000 0001 0010 0011 0100 0101 0110 0111 1000 1001 1010 1011 1100 1101 1110 1111
0000 0001 0010 0011 0100 0101 0110 0111 1000 1001 1010 1011 1100 1101 1110 1111

Figure 15.3. Operation-code locator—Hexadecimal 8080 operation-code matrix.

tinues from codes 4 to B, and section III continues from C to F. Section A extends from codes 0 to 7, and Section B extends from 8 to F.

Sections IA and IB contain the majority of new instructions related to the 8080 chip. These instructions should be read very carefully and fully understood since the INX instruction, as contrasted to the 8008 INC instruction, actually involves the incrementing of a register pair rather than merely a single register. In general, however, section I deals with the increment/decrement instructions and load instructions. Section II involves the movement of data between registers (code $4x$ to $7x$) and accumulator arithmetic and logic functions (codes $8x$ to $13x$). Section III completes the CALL, JUMP, and RETURN instructions, PUSH/POP instructions, RESET instruction, and a few new 8080 instructions.

Section A deals primarily with the B, D, H, and M registers, as well as the accumulator instruction codes for ADD, SUB, AND, and ORA. The "0" set flag instruction set for all the flag flip-flops rounds out the grouping.

Section B deals primarily with the C, E, L, and A registers as well as the accumulator instruction codes for ADC, SBB, XRA, and CMP. The "1" set flag instruction set for all the flag flip-flops rounds out the grouping. For the use of the Intel 8080 Hexadecimal Code Matrix, refer to the basic 8008 discussion in Chapter 12.

THE MOTOROLA M6800 MICROPROCESSOR SYSTEM

The Motorola M6800 Microprocessor is in many respects the most *vendor-oriented* system of the three microprocessor chips examined (Intel 8008, 8080, and Motorola M6800). There exists a *total product offering,* as Motorola states, of the MPU (microprocessor), ROM (1024×8 read only memory), RAM (128×8 random access memory), and PIA (peripheral interface adapter). With these four components a basic system can be configured. The M6800 chip, like the 8080 chip, comes in a 40-pin package. It is powered, however, by only a single $+5$-V supply, which makes the M6800 CPU an extremely convenient device.

The M6800 Program-instruction Set

The program instruction set of the Motorola M6800 appears to be very different from those of either the Intel 8008 or 8080 microprocessor until it is understood that the genius of the M6800 instruction codes lies in its seven addressing modes. It is within these seven addressing modes that the apparent lack of on-chip registers is compensated for. For this reason the seven modes of addressing are evaluated prior to providing the instruction set in detail.

Accumulator (ACCX) Addressing

The M6800 utilizes two accumulators to manipulate data, accumulators A and B, either of which can be addressed.

Immediate Addressing (2 bytes)

In immediate addressing, two bytes are generally used. The second byte contains the data to be operated on. Thus, to add the number 25 to accumulator *A* the program would be written:

> *XX*
> 8*B*
> 25   8B ≡ ADD immediate to *A*
> *XX*

The notable exceptions to the data being contained in the second byte are the LDX (load-index register) and LDS (load-stack pointer) instructions. Since these registers in the M6800 MPU are 16-bit registers, the immediate addressing instruction utilizes three bytes, an instruction byte followed by 2 data bytes.

Direct Addressing (2 bytes)

In direct addressing, two instruction bytes are used. The first byte contains the operation code followed by the second byte, which is the address of the data to be operated on. By this means any of the 256 bytes located on page 00 may be addressed. The major advantage of direct addressing is the capability of entering into memory using only a single byte of address, thus saving execution time, since a second (higher order address) byte does not need to be read, sent out on the address bus, and read by the memory device. An example of this program might be the need to enter a variable cost (Dow Jones Average) into a stock-market calculation. Since it changes every day, it will be placed into memory location 00,08. Any reference to the Dow Jones Average can be addressed to this location. Thus, to add the Dow Jones Average to the sum already in Accumulator *B*, the program would read:

> *XX*
> DB   ADD direct to *B*
> 08   (the)
> *XX*   Data in location 00,08

Extended Addressing (3 bytes)

Extended addressing is identical to direct addressing except that it uses three bytes of instruction. The address of the data to be operated on is contained in the second and third bytes of the instruction code. Thus, extended addressing increases the range of data locations from the first 256 memory locations to the entire 65,536 memory locations.

Indexed Addressing (2 bytes)

In indexed addressing, two bytes of instruction are used. While the instruction itself is similar to direct addressing, the address contained in the second byte of the instruction is added to the lower-order address of the index register (*a* 16-bit register) and any overflow is added to the higher-order address of the index register. Thus, if the index register is established as a touchstone in a given control loop, all data values for that control loop can be stored or retrieved based on that initial index-register value. A second control loop, operated by the same program, could be simultaneously operated using a separate block in memory by the simple modification of the index-register value. Assume the A register is to be compared with the F8th number stored in the control loop 2, whose index-register value is A0,44. The program for this instruction would be:

> XX
> AI Compare with accumulator A
> $F8$ index register + F8
> XX

The index register value is A0,44. Adding F8 to A0,44 yields A1, $3C$, the overflow from the addition of $F + 4 = 3 +$ an overflow of 1, adding one to the higher order address. The accumulator A would be compared with the number stored in memory location A1, $3C$.

Inherent Addressing (1 byte)

Inherent addressing utilizes a single byte. The operation code itself denotes the address. These include codes such as INC A (increase accumulator A), TST A (test accumulator A for 00 value), SBA (subtract accumulator B from accumulator A), or CLRB (clear accumulator B).

Relative Addressing (2 bytes)

Relative addressing is a 2-byte code and is only used in branching instructions. It is similar to the indexed addressing technique except that the address of the second byte is added to the program counter's lowest 8 bits (PC = 16 bit register) plus 2 (to avoid any existing 3-byte instruction codes) and the carry/borrow occurring used to adjust the higher order 8 bits of the program counter. This permits a block of memory to be branched to in a single byte within a range of -125 and $+129$ bytes around the present program counter. This instruction is used to increase the speed of the M6800 system.

Assume the program counter is presently set at 08,4F. Assume further that the program to be branched to (if $Z > 0$) is located at 09,02. The location 09,02 is B3 (113_{10}) larger than 08,4F. Thus, the relative addressing program for branching control to location 09,02 would be:

> XX
> $2E$ Branch relative if $Z > 0$
> $B3$ relative value = B3
> XX

Once the addressing modes of the M6800 MPU are understood, the remainder of the instruction set is straightforward.

The M6800 instruction set is shown in Figure 15.4. (Pay particular attention to the listing of the Boolean/arithmetic operation column to determine the physical operation being performed.)

In general, the operations indicated in the M6800 instruction set are similar to those operations already examined for the Intel devices. It should be noted, however, that the branch and jump instructions are identical except that the former always uses relative addressing restricted to a range of -125 to $+129$ bytes around the present program-counter location, whereas the jump instruction uses either indexed or extended addressing providing a range of transfer throughout the entire memory.

The only other new element in the M6800 instruction set is the condition-code register, which corresponds to the flag flip-flops in the Intel units. This register is organized as follows:

b_5	b_4	b_3	b_2	b_1	b_0
H	I	N	Z	V	C

H Half-carry; set whenever a carry from D_3 to D_4 of the resultant byte is generated by ADD, ABA, ADC. It is cleared by the absence of a D_3 to D_4 carry; not affected by other instructions.

I Interrupt mask; it is set by hardware or software interrupt or SEI instruction and is cleared by the CLI instruction. It is rarely used in arithmetic operations.

N Negative; it is set to 1 if value of high order bit D_7 is 1, otherwise, is set to 0.

Z Zero; if operation results equal 0, $Z = 1$. If the operation results $\neq 0$, $Z = 0$.

V Overflow; if an operation results in an overflow, $V = 1$. If no overflow occurs, $V = 0$.

C Carry; if the operation results in a carry from D_7, C $= 1$. If there is no carry from D_7, C $= 0$. (*Note:* in general, the D_7 bit is used to designate the negative number if 1, the positive number if 0.)

The Hexadecimal M6800 Operation Code Matrix

The M6800 Hexadecimal Operation Code Matrix is shown in Figure 15.5. Figure 15.6 illustrates the general location of the instruction groups. The matrix can be broken into horizontal sections as shown. Code section $2X$ provides all relative branching instructions. Code Section $4X$ provides miscellaneous operations on Accumulator A, and section $5X$ provides the same instruction set for accumulator B. Code sections $6X$ and $7X$ provide these instruction codes for indexed addressing and extended addressing, respectively.

Sections $8X$—BX provide the immediate, direct, extended, and indexed addressing codes for accumulator A, and sections CX—FX provide the identical codes for accumulator B.

For information on the use of the Hexadecimal M6800 Operation Code Matrix, refer to the Intel 8008 Hexadecimal Operation Code Matrix discussion in Chapter 12.

ADDRESSING MODES

ACCUMULATOR AND MEMORY OPERATIONS

OPERATIONS	MNEMONIC	IMMED OP	IMMED ~	IMMED #	DIRECT OP	DIRECT ~	DIRECT #	INDEX OP	INDEX ~	INDEX #	EXTND OP	EXTND ~	EXTND #	INHER OP	INHER ~	INHER #	BOOLEAN/ARITHMETIC OPERATION (All register labels refer to contents)	H (5)	I (4)	N (3)	Z (2)	V (1)	C (0)
Add	ADDA	8B	2	2	9B	3	2	AB	5	2	BB	4	3				$A + M \rightarrow A$	↕	•	↕	↕	↕	↕
	ADDB	CB	2	2	DB	3	2	EB	5	2	FB	4	3				$B + M \rightarrow B$	↕	•	↕	↕	↕	↕
Add Acmltrs	ABA													1B	2	1	$A + B \rightarrow A$	↕	•	↕	↕	↕	↕
Add with Carry	ADCA	89	2	2	99	3	2	A9	5	2	B9	4	3				$A + M + C \rightarrow A$	↕	•	↕	↕	↕	↕
	ADCB	C9	2	2	D9	3	2	E9	5	2	F9	4	3				$B + M + C \rightarrow B$	↕	•	↕	↕	↕	↕
And	ANDA	84	2	2	94	3	2	A4	5	2	B4	4	3				$A \bullet M \rightarrow A$	•	•	↕	↕	R	•
	ANDB	C4	2	2	D4	3	2	E4	5	2	F4	4	3				$B \bullet M \rightarrow B$	•	•	↕	↕	R	•
Bit Test	BITA	85	2	2	95	3	2	A5	5	2	B5	4	3				$A \bullet M$	•	•	↕	↕	R	•
	BITB	C5	2	2	D5	3	2	E5	5	2	F5	4	3				$B \bullet M$	•	•	↕	↕	R	•
Clear	CLR							6F	7	2	7F	6	3				$00 \rightarrow M$	•	•	R	S	R	R
	CLRA													4F	2	1	$00 \rightarrow A$	•	•	R	S	R	R
	CLRB													5F	2	1	$00 \rightarrow B$	•	•	R	S	R	R
Compare	CMPA	81	2	2	91	3	2	A1	5	2	B1	4	3				$A - M$	•	•	↕	↕	↕	↕
	CMPB	C1	2	2	D1	3	2	E1	5	2	F1	4	3				$B - M$	•	•	↕	↕	↕	↕
Compare Acmltrs	CBA													11	2	1	$A - B$	•	•	↕	↕	↕	↕
Complement, 1's	COM							63	7	2	73	6	3				$\overline{M} \rightarrow M$	•	•	↕	↕	R	S
	COMA													43	2	1	$\overline{A} \rightarrow A$	•	•	↕	↕	R	S
	COMB													53	2	1	$\overline{B} \rightarrow B$	•	•	↕	↕	R	S
Complement, 2's (Negate)	NEG							60	7	2	70	6	3				$00 - M \rightarrow M$	•	•	↕	↕	①	②
	NEGA													40	2	1	$00 - A \rightarrow A$	•	•	↕	↕	①	②
	NEGB													50	2	1	$00 - B \rightarrow B$	•	•	↕	↕	①	②
Decimal Adjust, A	DAA													19	2	1	Converts Binary Add. of BCD Characters into BCD Format	•	•	↕	↕	•	③
Decrement	DEC							6A	7	2	7A	6	3				$M - 1 \rightarrow M$	•	•	↕	↕	④	•
	DECA													4A	2	1	$A - 1 \rightarrow A$	•	•	↕	↕	④	•
	DECB													5A	2	1	$B - 1 \rightarrow B$	•	•	↕	↕	④	•
Exclusive OR	EORA	88	2	2	98	3	2	A8	5	2	B8	4	3				$A \oplus M \rightarrow A$	•	•	↕	↕	R	•
	EORB	C8	2	2	D8	3	2	E8	5	2	F8	4	3				$B \oplus M \rightarrow B$	•	•	↕	↕	R	•
Increment	INC							6C	7	2	7C	6	3				$M + 1 \rightarrow M$	•	•	↕	↕	⑤	•
	INCA													4C	2	1	$A + 1 \rightarrow A$	•	•	↕	↕	⑤	•
	INCB													5C	2	1	$B + 1 \rightarrow B$	•	•	↕	↕	⑤	•

COND. CODE REG.

Figure 15.4. M6800 Instruction set. (Courtesy of Motorola Semiconductor Products, Inc.)

Operation	Mnemonic	IMM OP	IMM ~	IMM #	DIR OP	DIR ~	DIR #	IND OP	IND ~	IND #	EXT OP	EXT ~	EXT #	INH OP	INH ~	INH #	Boolean/Arithmetic Operation
Load Acmltr	LDAA	86	2	2	96	3	2	A6	5	2	B6	4	3				$M \to A$
	LDAB	C6	2	2	D6	3	2	E6	5	2	F6	4	3				$M \to B$
Or, Inclusive	ORAA	8A	2	2	9A	3	2	AA	5	2	BA	4	3				$A + M \to A$
	ORAB	CA	2	2	DA	3	2	EA	5	2	FA	4	3				$B + M \to B$
Push Data	PSHA													36	4	1	$A \to M_{SP}, SP-1 \to SP$
	PSHB													37	4	1	$B \to M_{SP}, SP-1 \to SP$
Pull Data	PULA													32	4	1	$SP+1 \to SP, M_{SP} \to A$
	PULB													33	4	1	$SP+1 \to SP, M_{SP} \to B$
Rotate Left	ROL							69	7	2	79	6	3				M }
	ROLA													49	2	1	A } rotate left through carry
	ROLB													59	2	1	B }
Rotate Right	ROR							66	7	2	76	6	3				M }
	RORA													46	2	1	A } rotate right through carry
	RORB													56	2	1	B }
Shift Left, Arithmetic	ASL							68	7	2	78	6	3				M }
	ASLA													48	2	1	A } shift left, $0 \to b_0$
	ASLB													58	2	1	B }
Shift Right, Arithmetic	ASR							67	7	2	77	6	3				M }
	ASRA													47	2	1	A } shift right, arithmetic
	ASRB													57	2	1	B }
Shift Right, Logic.	LSR							64	7	2	74	6	3				M }
	LSRA													44	2	1	A } $0 \to b_7$ shift right
	LSRB													54	2	1	B }
Store Acmltr.	STAA				97	4	2	A7	6	2	B7	5	3				$A \to M$
	STAB				D7	4	2	E7	6	2	F7	5	3				$B \to M$
Subtract	SUBA	80	2	2	90	3	2	A0	5	2	B0	4	3				$A - M \to A$
	SUBB	C0	2	2	D0	3	2	E0	5	2	F0	4	3				$B - M \to B$
Subtract Acmltrs.	SBA													10	2	1	$A - B \to A$
Subtr. with Carry	SBCA	82	2	2	92	3	2	A2	5	2	B2	4	3				$A - M - C \to A$
	SBCB	C2	2	2	D2	3	2	E2	5	2	F2	4	3				$B - M - C \to B$
Transfer Acmltrs	TAB													16	2	1	$A \to B$
	TBA													17	2	1	$B \to A$
Test, Zero or Minus	TST							6D	7	2	7D	6	3				$M - 00$
	TSTA													4D	2	1	$A - 00$
	TSTB													5D	2	1	$B - 00$

Figure 15.4. M6800 Instruction set. (Courtesy of Motorola Semiconductor Products, Inc.)

INDEX REGISTER AND STACK POINTER OPERATIONS

POINTER OPERATIONS	MNEMONIC	IMMED OP	IMMED ~	IMMED #	DIRECT OP	DIRECT ~	DIRECT #	INDEX OP	INDEX ~	INDEX #	EXTND OP	EXTND ~	EXTND #	INHER OP	INHER ~	INHER #	BOOLEAN/ARITHMETIC OPERATION	H (5)	I (4)	N (3)	Z (2)	V (1)	C (0)
Compare Index Reg	CPX	8C	3	3	9C	4	2	AC	6	2	BC	5	3				$(X_H/X_L) - (M/M+1)$	•	•	⑦	↕	⑧	•
Decrement Index Reg	DEX													09	4	1	$X - 1 \rightarrow X$	•	•	↕	↕	•	•
Decrement Stack Pntr	DES													34	4	1	$SP - 1 \rightarrow SP$	•	•	•	•	•	•
Increment Index Reg	INX													08	4	1	$X + 1 \rightarrow X$	•	•	↕	↕	•	•
Increment Stack Pntr	INS													31	4	1	$SP + 1 \rightarrow SP$	•	•	•	•	•	•
Load Index Reg	LDX	CE	3	3	DE	4	2	EE	6	2	FE	5	3				$M \rightarrow X_H, (M+1) \rightarrow X_L$	•	•	⑨	↕	R	•
Load Stack Pntr	LDS	8E	3	3	9E	4	2	AE	6	2	BE	5	3				$M \rightarrow SP_H, (M+1) \rightarrow SP_L$	•	•	⑨	↕	R	•
Store Index Reg	STX				DF	5	2	EF	7	2	FF	6	3				$X_H \rightarrow M, X_L \rightarrow (M+1)$	•	•	⑨	↕	R	•
Store Stack Pntr	STS				9F	5	2	AF	7	2	BF	6	3				$SP_H \rightarrow M, SP_L \rightarrow (M+1)$	•	•	⑨	↕	R	•
Indx Reg → Stack Pntr	TXS													35	4	1	$X - 1 \rightarrow SP$	•	•	•	•	•	•
Stack Pntr → Indx Reg	TSX													30	4	1	$SP + 1 \rightarrow X$	•	•	•	•	•	•

JUMP AND BRANCH OPERATIONS

OPERATIONS	MNEMONIC	RELATIVE OP	RELATIVE ~	RELATIVE #	INDEX OP	INDEX ~	INDEX #	EXTND OP	EXTND ~	EXTND #	INHER OP	INHER ~	INHER #	BRANCH TEST	H (5)	I (4)	N (3)	Z (2)	V (1)	C (0)
Branch Always	BRA	20	4	2										None	•	•	•	•	•	•
Branch If Carry Clear	BCC	24	4	2										C = 0	•	•	•	•	•	•
Branch If Carry Set	BCS	25	4	2										C = 1	•	•	•	•	•	•
Branch If = Zero	BEQ	27	4	2										Z = 1	•	•	•	•	•	•
Branch If ≥ Zero	BGE	2C	4	2										$N \oplus V = 0$	•	•	•	•	•	•
Branch If > Zero	BGT	2E	4	2										$Z + (N \oplus V) = 0$	•	•	•	•	•	•
Branch If Higher	BHI	22	4	2										$C + Z = 0$	•	•	•	•	•	•
Branch If ≤ Zero	BLE	2F	4	2										$Z + (N \oplus V) = 1$	•	•	•	•	•	•
Branch If Lower Or Same	BLS	23	4	2										$C + Z = 1$	•	•	•	•	•	•
Branch If < Zero	BLT	2D	4	2										$N \oplus V = 1$	•	•	•	•	•	•
Branch If Minus	BMI	2B	4	2										N = 1	•	•	•	•	•	•
Branch If Not Equal Zero	BNE	26	4	2										Z = 0	•	•	•	•	•	•
Branch If Overflow Clear	BVC	28	4	2										V = 0	•	•	•	•	•	•
Branch If Overflow Set	BVS	29	4	2										V = 1	•	•	•	•	•	•
Branch If Plus	BPL	2A	4	2										N = 0	•	•	•	•	•	•
Branch To Subroutine	BSR	8D	8	2										} See Special Operations	•	•	•	•	•	•
Jump	JMP				6E	4	2	7E	3	3				} See Special Operations	•	•	•	•	•	•
Jump To Subroutine	JSR				AD	8	2	BD	9	3					•	•	•	•	•	•
No Operation	NOP										01	2	1	Advances Prog. Cntr. Only	•	•	•	•	•	•
Return From Interrupt	RTI										3B	10	1		⑩	⑩	⑩	⑩	⑩	⑩
Return From Subroutine	RTS										39	5	1	} See special Operations	•	•	•	•	•	•
Software Interrupt	SWI										3F	12	1	} See special Operations	•	•	•	•	•	•
Wait for Interrupt	WAI										3E	9	1		•	S ⑪	•	•	•	•

CONDITIONS CODE REGISTER

OPERATIONS	MNEMONIC	INHER OP	~	#	BOOLEAN OPERATION	5 H	4 I	3 N	2 Z	1 V	0 C
Clear Carry	CLC	0C	2	1	$0 \rightarrow C$	•	•	•	•	•	R
Clear Interrupt Mask	CLI	0E	2	1	$0 \rightarrow I$	•	R	•	•	•	•
Clear Overflow	CLV	0A	2	1	$0 \rightarrow V$	•	•	•	•	R	•
Set Carry	SEC	0D	2	1	$1 \rightarrow C$	•	•	•	•	•	S
Set Interrupt Mask	SEI	0F	2	1	$1 \rightarrow I$	•	S	•	•	•	•
Set Overflow	SEV	0B	2	1	$1 \rightarrow V$	•	•	•	•	S	•
Acmltr A → CCR	TAP	06	2	1	$A \rightarrow CCR$	—	—	—	(12)	—	—
CCR → Acmltr A	TPA	07	2	1	$CCR \rightarrow A$	•	•	•	•	•	•

CONDITION CODE REGISTER NOTES:

(Bit set if test is true and cleared otherwise)

1. (Bit V) Test: Result = 10000000?
2. (Bit C) Test: Result = 00000000?
3. (Bit C) Test: Decimal value of most significant BCD Character greater than nine? (Not cleared if previously set.)
4. (Bit V) Test: Operand = 10000000 prior to execution?
5. (Bit V) Test: Operand = 01111111 prior to execution?
6. (Bit V) Test: Set equal to result of $N \oplus C$ after shift has occurred.
7. (Bit N) Test: Sign bit of most significant (MS) byte of result = 1?
8. (Bit V) Test: 2's complement overflow from subtraction of LS bytes?
9. (Bit N) Test: Result less than zero? (Bit 15 = 1)
10. (All) Load Condition Code Register from Stack. (See Special Operations)
11. (Bit I) Set when interrupt occurs. If previously set, a Non-Maskable Interrupt is required to exit the wait state.
12. (ALL) Set according to the contents of Accumulator A.

LEGEND:

OP Operation Code (Hexadecimal);
~   Number of MPU Cycles;
#   Number of Program Bytes;
+   Arithmetic Plus;
−   Arithmetic Minus;
•   Boolean AND;
Msp Contents of memory location pointed to be Stack Pointer;
+   Boolean Inclusive OR;
⊕   Boolean Exclusive OR;
M̄   Complement of M;
→   Transfer Into;
0   Bit = Zero;

00 Byte = Zero;
H   Half-carry from bit 3;
I   Interrupt mask
N   Negative (sign bit)
Z   Zero (byte)
V   Overflow, 2's complement
C   Carry from bit 7
R   Reset Always
S   Set Always
‡   Test and set if true, cleared otherwise
•   Not Affected
CCR Condition Code Register
LS Least Significant
MS Most Significant

Figure 15.4 (cont.) M6800 Instruction set. (Courtesy of Motorola Semiconductor Products, Inc.)

223

Second digit

First digit	0	1	2	3	4	5	6	7	8	9	(10) A	(11) B	(12) C	(13) D	(14) E	(15) F
0																
1				Miscellaneous codes												
2				Relative branching codes												
3																
4				Accumulator A codes												
5				Accumulator B codes												
6				Indexed addressing												
7				Extended addressing												
8				Accumulator A												
9				Immediate addressing												
(10) A				Direct addressing												
(11) B				Extended addressing Indexed addressing												
(12) C				Accumulator B												
(13) D				Immediate addressing												
(14) E				Direct addressing												
(15) F				Extended addressing Indexed addressing												

Binary equivalents:

Hex	Binary
0	0000
1	0001
2	0010
3	0011
4	0100
5	0101
6	0110
7	0111
8	1000
9	1001
(10) A	1010
(11) B	1011
(12) C	1100
(13) D	1101
(14) E	1110
(15) F	1111

Figure 15.6. Operation code locator—hexadecimal M6800 operation-code matrix.

PROBLEMS

1. At a convention a door prize is to be given away at uniform intervals. These intervals correspond to the sum of 250 computer units as determined for the individuals entering the convention. A computer unit is determined by the running unit count as follows. The entrant's age minus the years he has worked at his present job or his age divided by 2, whichever is less. When the running count reaches *exactly* 250 computer units, the individual entering wins the door prize. If the running count *exceeds* 250, the prize will be awarded to the next individual entering the convention. The winner will be notified by means of a flashing red light.

   Develop the flowcharts and the 8080 program to perform this task. The person's age will be inputted at In 1 and the years worked at the present job at In 2. The output signifying a winner will be at Out 15.

2. Repeat problem 1, except use the M6800 program to perform this task.

3. Develop an 8080 program to provide change from 1¢ to 99¢ with the fewest number of coins possible. The available coins will be inputted as follows: 1¢ → In 1, 5¢ → In 2, 10¢ → In 3, 25¢ → In 4, and 50¢ → In 5. The corresponding change will be outputted as follows: 1¢ → Out 12, 5¢ → Out 13, 10¢ → Out 14, 25¢ → Out 15, and 50¢ → Out 16. The value of the change required will be inputted at In 6.

4. Repeat problem 3 except use the M6800 program to perform the task.

5. Two trains travel on separate tracks until they reach Middletown. The trains share a common track from Middletown to Hartford some 30 mi apart. The tracks divide again at Hartford. The trains can be traveling at a speed between 30 m.p.h. and 60 m.p.h., depending on the cargo they are carrying. Under *no* circumstance is train A to stop. Develop the flowcharts and the 8080 computer program to determine if train B will be required to stop.

   Sensors A and B are set to determine exact train velocity in m.p.h. Sensor C responds to the presence or absence of a train on the common track. If train B is stopped, it *must* be started again as soon as it is safe. Develop the 8080 flowcharts and machine program to perform this task.

6. Repeat problem 5 except use the M6800 program to perform the task.

APPENDIX

Glossary of Microprocessor Terms

Access Time. The time period required for the computer to move data from the CPU to memory or memory to CPU.

Address. The number used by the computer to designate the location of memory data or input/output registers.

ALU. See arithmetic logic unit.

Analog-to-Digital Converter (A/DC). An interface circuit which is used to convert an analog signal into a digital signal.

Architecture. Organizational structure of a computer system.

Arithmetic logic unit (ALU). The portion of the central processing unit (CPU) that performs the arithmetic and logical operations for the system.

Assembler. The software program that converts assembly-language operation codes into machine-language operation codes for use by the computer.

Assembly Language. Generally, a group of alphabet characters or symbols called *mnemonics* used as substitutes for machine-language numeric instructions. It is often felt that a mnemonic is easier to remember than the numeric machine code; hence, programming is enhanced.

Binary. A number system consisting of only two digits, 0 and 1. This is in contrast to the use of 10 digits (0–9) as defined in the decimal system. In the computer the use of "digital," "binary," and "two-state" are essentially synonymous.

Bit. A bit is a single binary digit. A bit can have either a 1 or 0 value.

Bit Slice. A multichip microprocessor that has been designed to operate with a single control section and one or more identical arithmetic and logic units (ALU) and register/memory units in parallel. For example, three 4-bit slices connected in parallel with a single control chip would produce a 12-bit microprocessor.

Bus. A group of dedicated wires performing a common task. In general, a maximum of three bus systems are used for the microprocessor system: the address bus, the data bus, and the input/output bus. These three bus systems may be a single set of wires that are time shared.

Byte. Generally used, a byte is the number of bits that the computer processes as a unit. The byte may be different than the computer word length (bits in a word). For example, a 16-bit word length computer and an 8-bit microprocessor may both perform data operations on 8-bit bytes.

Central Processing Unit. The portion of the computer that controls the execution of operation instructions. The CPU usually contains the ALU, timing and control unit, temporary memory storage, program counter/address stack, instruction registers and I/O, and data and memory bus interface amplifiers.

Compiler. The subunit that is used to convert programs written in high-level programming languages into the machine-operation codes used by the computer.

CPU. See central processing unit.

Cycle Time. The time interval required to perform a fundamental sequence of operations.

Digital-to-analog Converter (DAC). The interface circuit used to convert the output digital byte into an analog signal.

Direct Memory Access. A computer mechanism that allows the input/output devices to control the operation of the computer in order to directly write into or read out of memory without passing through the accumulator.

DMA. See direct memory access.

EAROM. Electrically alterable ROM. A ROM that can be erased and reprogrammed any number of times. This term is generally used for ROMs that can be altered without being removed from the memory circuit.

Hardware. Denotes a group of solid-state chips mounted on printed-circuit modules used to perform electronic functions.

Hexadecimal. A number system utilizing a total of 16 digits (0–9, A,B,C,D,E,F) that is used to interpret a 4-bit binary-code format.

Higher-level Language. Program language derived from user-oriented functions. Fortran, Cobol, Basic, and PL/1 are four higher-level languages in current use. A single statement in higher-level language program may translate into a major subroutine within the machine-level program of the computer. The higher-level language requires the use of a compiler to interface with the machine being programmed.

Input–Output. Applies to the input/output registers of the computer. However, due to its application to large computer installations, the term has become synonymous with the peripheral equipment used to communicate with the computer.

Instruction. A group of bits used to define a computer operation. An instruction may move data, cause arithmetic or logic operations to be performed, control I/O, or result in machine decisions.

Instruction Set. The total complement of instructions that a computer can perform. In general, each computer has a different instruction set. Often several instructions within an instruction set may perform a nearly identical task. Thus, the instruction set cannot be used to indicate the quality of a computer.

Interrupt. A signal to the computer that a subroutine, not included in the normal routine of the computer, is to be performed. The computer suspends normal operation, performs the desired subroutine, and then returns to normal operation. Interrupts are generally used to introduce nonsynchronous input data to the computer.

Large-scale Integration (LSI). High-density monolithic integrated circuits having the equivalent of several hundred semiconductor devices diffused into a single chip.

LSI. See large-scale integration.

Machine Cycle. The basic machine level sequence performed by the computer. In general, the cycle sequence is:

1. Read the operation to be performed (fetch cycle).
2. Establish the next step to be performed (increment program counter).
3. Set up the required internal gates to perform the instruction.
4. Perform the operation.

This sequence is continually repeated until the computer is issued a halt instruction.

Machine Language.   A set of numeric codes that control the functioning of the computer. These codes are maintained in the program memory and establish the logic gating of the computer's logic circuits.

Mask-programmable ROM.   The technique of vendor modification of the ROM bit pattern during the final phase of device fabrication. The final device interconnect mask is constructed to the customer's needs and the final interconnect pattern is diffused onto the chip to create the desired ROM program. This form of ROM programming can only be performed at the vendor's facility.

Medium-scale Integration.   Usually refers to a monolithic chip device containing more than 20 semiconductors and less than several hundred semiconductors on a single chip. The exact transition from medium- to large-scale integration is very vague.

Memory.   That part of the computer holding data and instructions. Memory can be maintained as temporary or permanent memory. Each data and instruction word is provided a unique location (address) in memory.

Microprocessor.   A computer whose major elements (CPU, ALU, and memory) are contained on a single chip or a few chips at most.

Microprogram.   A program in memory that is initiated by a single instruction located in the main memory. The appellation does not have reference to a program as used in a microprocessor but rather to its use as a portion of the program as a whole.

Mnemonic Code.   A technique used to create an instruction code that is easy to remember. This code is usually associated with the assembly language that requires conversion to the machine-level program.

MSI.   See medium-scale integration.

Octal.   A number system utilizing a total of eight digits (0–7) that is used to interpret a 3-bit binary-code format.

Operation Code.   See instruction. These two terms are synonymous.

Phase-locked Loop (PLL).   A series of analog or digital devices acting synergistically to form an electronic closed-loop servo system. The PLL can be used to demodulate a frequency-modulation signal into an analog signal.

PLA.   See programmed-logic array.

PLL.   See phase-locked loop.

Programmed-logic Array (PLA).   An LSI chip which can be programmed to represent a series of AND gates and OR gates. This device can replace a large number of SSI digital chips by a single chip.

Program.   A set of coded instructions used to direct the computer to perform a desired set of operations or yield a desired solution to a specific program.

PROM.   A programmable ROM. This ROM can be programmed by the user once, and thereafter becomes a ROM.

RAM.   See random-access memory.

Random-access Memory (RAM).   Usually referred to as read/write memory that can be addressed instantaneously; that is, any bit within memory can be accessed in the same time as any other bit.

Read-only Memory (ROM).   A memory that holds permanent data or instructions. A ROM cannot be written into by the user, and its memories are used to store programs, mathematical tables, or continually used permanent data.

Register.   A storage circuit that can be quickly accessed (normally in parallel) and the data temporarily "locked in." Registers are used for temporary storage of data.

ROM.   See read-only memory.

Scratch-pad Memory.   Random-access memory used for temporary storage of data or intermediate results.

Sequential Memory.   Memory in which the data must be read in sequence (serial). Since the data must be read out of memory in the same order as written into memory, access time to reach a desired piece of data may be long. A tape recording of data is an example of sequential memory.

Small-scale Integration (SSI).   Usually refers to monolithic chips of the 5400/ 7400 series containing several semiconductor devices on a single chip.

Software.   Coded instructions to direct the operation of the computer. In general, the program and software are synonymous.

SSI.   See small-scale integration.

Stack.   A sequence of registers or memory locations used in a LIFO fashion (last in–first out). A stack pointer specifies the last entry into the stack.

Stack Pointer.   A counter used to locate a stack in memory.

Volatile Memory.   A memory whose contents are lost when operating power is removed.

Word.   The number of bits needed to represent the largest data element normally processed by the computer.

REFERENCES

Bracketed chapters denote chapter in this book where referenced text was useful.

1. Luecke, Mize, and Carr, *Semiconductor Memory Design and Application,* McGraw-Hill, New York, 1973. [Chapters 3, 5, 6, 7.]

2. Engineering Staff of American Micro-Systems, Inc., *MOS Integrated Circuits,* Penney and Lau, Eds., Von Nostrand Reinhold, New York, 1972. [Chapters 3, 5, 6, 7.]

3. Users Manual—8008 8-bit Parallel Central Processor Unit, Rev. 4, Intel Corporation, Santa Clara, Calif., Nov. 1973. [Chapters 2, 5, 6, 7, 8, 9, 10, 11.]

4. Users Manual: Intel 8080 Microcomputer System Manual, 2nd printing, Intel Corporation, Santa Clara, Calif., Jan. 1975. [Chapter 12.]

5. M6800 Microprocessor Programming Manual, 2nd edition, Motorola Inc., Phoenix, Ariz., 1975. [Chapter 12.]

6. M6800 Microprocessor Applications Manual, 1st edition, Motorola Inc., Phoenix, Ariz., 1975. [Chapter 12.]

7. The Intel Memory Design Handbook, Intel Corporation, Santa Clara, Calif., Aug. 1973. [Chapters 5, 6, 7.]

8. Application Manual MF 8008, Bulletin 80007, Microsystems International, Ottawa, Canada, 1974. [Chapters 5, 6, 7.]

9. *Microcomputer Design* (2 vols.), Martin Research Ltd., Chicago, Ill., 1974. [Chapters 5, 6, 7, 9.]

10. D. D. McCracken, *Digital Computer Programming,* Wiley, New York, 1965.

11. Altman, Rosenblatt, Scrupski, Walker, Riley, and Riezenman, "Diverse Industry Users Clamber Aboard the Microprocessor Bandwagon," *Electronics,* McGraw-Hill, New York, July 11, 1974. [Chapter 2.]

12. Laurence Altman, "Single Chip Microprocessor Open Up a New World of Applications," *Electronics,* McGraw-Hill, New York, Apr. 18, 1974. [Chapter 12.]

13. Joel Altstein, "I²L: Today's Versatile Vehicle for Tomorrow's Custom LSI," *EDN,* Cahners, Chicago, Ill., Feb. 20, 1975. [Chapter 3.]

14. Darren Appelt, "Get Standby LSI Memory Power," *Electronic Design,* 12, Hayden, Rochelle Park, N. J., June 7, 1974. [Chapter 3.]

15. Larry Armstrong, "Microprocessors Steer to Detroit," *Electronics,* McGraw-Hill, New York, Apr. 18, 1974. [Chapter 12.]

16. Thomas Blakeslie, "Powering Up a Cartridge Tape Drive," *Digital Design,* Benwill, Brookline, Mass., Feb. 1975. [Chapter 9.]

17. John Bond, "Designers Guide to: Software for the Hardware Designer," Parts 1 and 2, *EDN,* Cahners, Chicago, Ill., Aug. 5, 1974. [Chapter 10.]

18. John Bond, "Designing Memories With 4K RAM's Will Be Easier Than With 1K Chips," *EDN,* Cahners, Chicago, Ill., July 5, 1974. [Chapter 6.]

19. Dick Brunner, "Sense Amps and Comparators," Motorola Semiconductors, Phoenix, Ariz., 1972. [Chapter 9.]

20. J. S. Byrd, "When Your System's Data Rates Differ, It Is Time For a Microprocessor," *EDN,* Cahners, Chicago, Ill., Nov. 20, 1974. [Chapters 10, 11.]

21. Callan and Baskin, "Microprocessor System Design," *Digital Design,* Benwill, Brookline, Mass., Feb. 1975. [Chapters 2, 11.]

22. Robert H. Cushman, "Understanding the Microprocessor Is No Trivial Task," *EDN, Cahners*, Chicago, Ill., Nov. 20, 1973. [Chapters 2, 11.]

23. Robert H. Cushman, "A Very Complete Chip Set Joins the Great Microprocessor Race," *EDN, Cahners*, Chicago, Ill., Nov. 20, 1974. [Chapters 2, 11, 12.]

24. Robert H. Cushman, "What Can You Do With a Microprocessor?," *EDN*, Cahners, Chicago, Ill., March 20, 1974. [Chapters 2, 11, 12.]

25. Robert H. Cushman, "Single Chip Microprocessor Move Into the 16-bit Arena," *EDN, Cahners*, Chicago, Ill., Feb. 20, 1975. [General.]

26. Robert H. Cushman, "Microprocessors Are Changing Your Future. Are You Prepared?," *EDN, Cahners*, Chicago, Ill., Nov. 5, 1973. [General.]

27. Robert H. Cushman, "Experts Predict The Future of Microprocessor Components," *EDN, Cahners*, Chicago, Ill., Feb. 20, 1974. [Chapters 3, 12.]

28. Robert H. Cushman, "Don't Overlook The 4-bit Up: They're Here and They're Cheap," *EDN, Cahners*, Chicago, Ill., Feb. 20, 1974. [Chapter 4.]

29. Sidney Davis, "Selection and Application of Semiconductor Memories," *Computer Design*, Computer Design, Concord, Mass., Jan. 1974. [Chapter 6.]

30. Sidney Davis, "A Fresh View of Mini and Microcomputers," *Computer Design*, Computer Design, Concord, Mass., May 1974. [Chapter 2.]

31. Sidney Davis, "Update On Magnetic Tape Memories," *Computer Design*, Computer Design, Concord, Mass., Aug. 1974. [Chapter 5.]

32. S. S. Durvasula, "Use 8 and 16K Sensing Techniques In Core Memory Designs," *EDN, Cahners*, Chicago, Ill., Oct. 5, 1973. [Chapter 5.]

33. Fred Etcheverry, "An Approach To Digital Recording By Low Cost Audio Cassette," *Computer Design*, Computer Design, Concord, Mass., Oct. 1974. [Chapter 5.]

34. James E. Fischer, "4K RAM's: Increased Densities Bring Difficult Testing Problems," *EDN, Cahners*, Chicago, Ill., Nov. 20, 1974. [Chapter 5.]

35. Mike Gerhufe, "More Bits/Chips Lead to Economical Semiconductor Memory Systems," *EDN, Cahners*, Chicago, Ill., Feb. 20, 1973. [Chapter 5.]

36. Ken Gorman, "The Programmable Logic Array: A New Approach To Microprogramming," *EDN, Cahners*, Chicago, Ill., Nov. 20, 1973. [Chapter 6.]

37. Marcian Hoff, Intel Corporation, "Designing With RAMs," Parts I and II., *EDN, Cahners*, Chicago, Ill., Aug. 1973. [Chapter 5.]

38. R. Holt and M. Lemas, "Current Microcomputer Architecture," *Computer Design*, Computer Design, Concord, Mass., Feb. 1974. [Chapters 2, 3, 5, 6, 7, 8, 9.]

39. George King, "Communication Systems in 1980 Will Provide 'Smart' Functions In Every Terminal," *Digital Design*, Benwill, Brookline, Mass., Nov. 1974. [Chapter 12.]

40. Joseph H. Kroeger, "Free Your Memory System From The Pause That Refreshes," *Electronic Products Magazine*, Garden City, New York, March 19, 1973. [Chapter 6.]

41. Donald R. Lewis, "Microprogramming: More 'In' Than Ever," *Electronic Design*, 17, Hayden, Rochelle Park, N. J., Aug. 16, 1973. [Chapters 10, 11.]

42. W. Lile, J. Scott, and Dr. A. Dingwall, "SOS CMOS Random Access Memories: A Mini Survey," *EDN, Cahners*, Chicago, Ill., Nov. 5, 1974. [Chapter 3.]

43. Robert H. F. Lloyd, "RAM Technology, MOS or Bipolar?," *Electronic Product Magazine*, Garden City, N. Y., June 18, 1973. [Chapter 3.]

44. Joe Maggiore, "PLA—A Universal Logic Element," *Electronic Product Magazine*, Garden City, N. Y., April 15, 1974. [Chapter 7.]

45. M. Maloney and S. Murahashi, "Design Considerations for Ruggedized Memories," *Computer Design*, Computer Design, Concord, Mass., May 1974. [Chapter 7.]

46. Joseph J. McDowell, "Improve ROM Systems With PROMs," *Electronic Design*, 14, Hayden, Rochelle Park, N. J., July 5, 1974. [Chapter 8.]

47. John McMullen, "Programming The PROM: A Make or Buy Decision," *EDN*, Cahners, Chicago, Ill., March 5, 1974. [Chapter 8.]

48. Al Moore, "Printer Control: A Minor Task For A Fast Microprocessor," *Electronic Design*, 25, Hayden, Rochelle Park, N. J., Dec. 6, 1974. [Chapter 9.]

49. Jerry Metzger, "The Year Of The 4K RAM," *Electronic Product Magazine*, Garden City, N. Y., March 18, 1974. [Chapter 6.]

50. Jerry Metzger, "Forum: It's Go For Microprocessors," *Electronic Product Magazine*, Garden City, N. Y., Nov. 19, 1973. [Chapter 12.]

51. David Mills, "Digital Cassettes—The Case For Convenience," *Electronic Product Magazine*, Garden City, N. Y., Oct. 15, 1973. [Chapter 9.]

52. Dale Mrozek, "PLAs Replace ROMs For Logic Designs," *Electronic Design*, 22, Hayden, Rochelle Park, N. J., Oct. 25, 1973. [Chapter 7.]

53. Jerry Neth, "Microprocessors and Microcomputers: What Will The Future Bring?," *EDN*, Cahners, Chicago, Ill., Nov. 20, 1974. [Chapters 2, 12.]

54. Jerry L. Ogdin, "Survey of 8-bit Microprocessors Reveals Wide Choice of Users," *EDN*, Cahners, Chicago, Ill., June 20, 1974. [General.]

55. Jerry L. Ogdin, "Survey of Microprocessors Reveals Limitless Variety," *EDN*, Cahners, Chicago, Ill., April 20, 1974. [General.]

56. William H. Ohm, "Reel-To-Reel Drive Design For A Cassette Recorder," *Computer Design*, Computer Design, Concord, Mass., Aug. 1973. [Chapter 9.]

57. Walt Patstone, "Brake Tester Takes Microprocessor For A Ride," *EDN*, Cahners, Chicago, Ill., Nov. 20, 1974. [Chapter 9.]

58. Tom Pittmar, "Improve Interrupt—Handling Capability Of Microprocessor With A Few ICs," *Electric Design*, 24, Hayden, Rochelle Park, N. J., Nov. 22, 1974. [Chapter 11.]

59. DuWayne A. Pople, "Keyboard Expand In Complexity," *Electronic Product Magazine*, Garden City, N. Y., April 15, 1974. [Chapter 9.]

60. Howard A. Raphael, "Join Micros Into Intelligent Networks," *Electronic Design*, 5, Hayden, Rochelle Park, N. J., March 1, 1975. [General.]

61. Douglas Risch, "Design D/A and A/D Interfaces For Your Computer," *EDN*, Cahners, Chicago, Ill., April 5, 1974. [Chapter 9.]

62. Stanley Runyon, "Focus on Keyboards," *Electronic Design*, 23, Hayden, Rochelle Park, N. J., Nov. 9, 1972. [Chapter 9.]

63. Roy R. Shanks, "Amorphous Semiconductors For Electrically Alterable Memory Application," *Computer Design*, Computer Design, Concord, Mass., May 1974. [Chapters 3, 7.]

64. M. Shima and F. Foggin, "In Switch To N-MOS Microprocessor Gets 2-μs Cycle Time," *Electronics*, McGraw-Hill, New York, Apr. 18, 1974. [Chapter 3.]

65. Lars Soderholm, "Microprocessors—Looking Ahead," *Design News*, Cahners, Chicago, Ill., Jan. 20, 1975. [General.]

66. Stern, "Microcomputers—Preparations For An Explosion," *Motorola Monitor*, Motorola Semiconductor Inc., Phoenix, Ariz., Dec. 1973. [General.]

67. L. Solomon and T. Swithenbank, "Floppy Disks," *Digital Design*, Benwill, Brookline, Mass., Nov. 1974. [Chapters 4, 5, 9.]

68. Gerald Strehl, "A User's Introduction To Rotating Memories," *Electronic Products Magazine*, Garden City, N. Y., March 19, 1973. [Chapters 4, 5.]

69. Jonathan Titus, "How To Design A μP-Based Controller System," *EDN*, Cahners, Chicago, Ill., Aug. 20, 1974. [General.]

70. Dave Uimari, "PROM's—A Practical Alternative To Random Logic," *Electronic Product Design*, Garden City, N. Y., Jan. 21, 1974. [Chapter 7.]

71. Alan J. Weissberger, "MOS/LSI Microprocessor Selection," *Electronic Design*, 12, Hayden, Rochelle Park, N. J., June 7, 1974. [General.]

72. Erwin Vodovoz, "Parallel Microprocessors Improve Hybrid Computer Performance," *EDN*, Cahners, Chicago, Ill., June 5, 1974. [General.]

73. Erwin Vodovoz, "N-Channel Microcomputers Need Less Interface," *Electronic Product Magazine*, Garden City, N. Y., June 17, 1974. [General.]

74. Alen Weissberger, "Distributed Function Microprocessor Architecture," *Computer Design,* Computer Design, Concord, Mass., Nov. 1974. [General.]

75. W. Wickes, "A Compatible MOS/LSI Microprocessor Device Family," *Computer Design,* Computer Design, Concord, Mass., July 1973. [General.]

76. David Wyland, "Using P/ROMs As Logic Elements," *Computer Design,* Computer Design, Concord, Mass., Sept. 1974. [Chapter 7.]

77. L. Young, T. Bennett, and J. Lovell, "*N*-Channel MOS Technology Yields New Generation Of Microprocessors," *Electronics,* McGraw-Hill, New York, Apr. 18, 1974. [Chapter 3.]

78. "Bipolar Microprocessor—The First To Hit The Market," *News Scope,* Electronic Design, Hayden, Rochelle Park, N. J., Sept. 13, 1974. [General.]

79. "Microprogramming Kit Allows User To Form Custom Microprocessors," Editorial, *Computer Design,* Computer Design, Concord, Mass., Aug. 1974. [General.]

80. "Programmable Logic Arrays: A Dormant Giant Awakening," Designers Roundtable, *EDN,* Cahners, Chicago, Ill., March 5, 1975. [Chapter 7.]

81. "Printers," *Digital Design,* Benwill, Brookline, Mass., Dec. 1974. [Chapters 5, 9.]

82. "Digital Tape Drive," *Digital Design,* Benwill, Brookline, Mass., Dec. 1974. [Chapters 5, 9.]

83. "Disk Storage Devices," *Digital Design,* Benwill, Brookline, Mass., Dec. 1974. [Chapters 5, 9.]

84. "Single Monolithic Chip Holds 16-bit Microprocessor," *New Products, Electronic Design,* Hayden, Rochelle Park, N. J., Dec. 6, 1974. [General.]

85. Jules Gilder, "MNOS Memory Upstaging MOS and Fixed Heads In Some Areas," Editorial, *Electronic Design,* Hayden, Rochelle Park, N. J., Sept. 1, 1973. [General.]

86. Daniel H. Sheingold, Ed., *Analog–Digital Conversion Handbook,* Analog Devices, Inc., Norwood, Mass., 1972. [Chapter 9.]

INDEX